EMBEDDED and NETWORKING SYSTEMS

Design, Software, and Implementation

T0239944

EMBEDDED and NETWORKING SYSTEMS

Design, Software, and Implementation

EDITED BY

GUL N. KHAN

KRZYSZTOF INIEWSKI

CRC Press
Taylor & Francis Group
Boca Raton   London   New York

CRC Press is an imprint of the
Taylor & Francis Group, an **informa** business

CRC Press
Taylor & Francis Group
6000 Broken Sound Parkway NW, Suite 300
Boca Raton, FL 33487-2742

First issued in paperback 2017

Version Date: 20130719

ISBN 13: 978-1-4665-9065-6 (hbk)
ISBN 13: 978-1-138-07405-7 (pbk)

Library of Congress Cataloging-in-Publication Data

Embedded and networking systems : design, software, and implementation / editors, Gul
   N. Khan, Krzysztof Iniewski.
       pages cm -- (Devices, circuits, and systems)
   Includes bibliographical references and index.
   ISBN 978-1-4665-9065-6 (hardcover : alk. paper)
   1. Embedded computer systems. I. Khan, Gul N. II. Iniewski, Krzysztof, 1960-

TK7895.E42E535 2014
006.2'2--dc23   2013027628

Visit the Taylor & Francis Web site at
http://www.taylorandfrancis.com

and the CRC Press Web site at
http://www.crcpress.com

Contents

Preface

Future embedded and network system designers will be more concerned with the application based tailoring of different elements of the system rather than concentrating on the details of system components such as processors, memories, routers, sensors, and other hardware modules of the system. These designers may have a surface knowledge of processors, sensors, routers, and other elements of the system. The success of twentieth century system designers depends on their capabilities and skills in dealing with the trade-offs to optimize system performance, cost, power, and other attributes relevant to system level non-functional requirements. In this book, we deal with the issues related to the design and synthesis of high performance embedded computer systems, particularly system-on-chip as well as mobile sensor and video networks encompassing embedded devices. The potential of embedded systems ranges from the simplicity of appliance control, consumer electronics, and sharing digital media to the coordination of a variety of complex joint actions carried out between collections of networked devices. Our book explores the emerging use of embedded systems and networking technologies from theoretical and practical applications. For embedded system design, a high performance embedded application is partitioned into various tasks that can be implemented on hardware or software running on the embedded processor. This book is organized to introduce issues related to the co-design of embedded systems, code optimization for a variety of embedded and networking applications, power and performance trade-offs, benchmarks for evaluating embedded systems and their components, and mobile sensor network systems. It also includes some novel applications such as mobile sensor systems and video networks.

The focus of our book is to provide a comprehensive review of both the groundbreaking technology as well as applications in the field of embedded and networking systems. The book attempts to expose the readers to historical evolution and trends as well as fundamental and analytical understanding of underlying technologies for embedded and networking systems. Embedded systems and devices are present all around us in the industrialized world. Embedded device research is being carried out in Asia, Europe, and North America. Looking at the origin of the contributions to this book, which comes from Canada, Austria, USA, Japan, Mexico, Singapore, and Korea, it can be said that this global aspect is well reflected. The main feature of this book lies in the timely presentation of latest research and development in the field and from sources all over the world, providing an interesting picture of the current state on a global scale.

The book contains a collection of chapters written by the researchers and experts of embedded code optimization, low-power embedded devices, co-synthesis of embedded systems, mobile sensor systems, and video networks. The emphasis is also on the fundamental concepts and analytical techniques that are applicable to a range of embedded and networking system applications rather than on specific embedded architectures, software development, applications, or system level integration.

Fully covering all these topics in one book is almost impossible; however, in-depth knowledge of all the topics is not necessarily required by an embedded system designer. A system designer can focus on embedded and networking systems from the system point of view: a basic knowledge of different components of the system. Our book encompasses a wide range of embedded and networking system technologies. The book can be divided into two sections, where Chapters 1 through 6 are the first section covering embedded processors and systems, software design, and implementation techniques. The second section of the book is related to networking systems including the sensor networks. Mobile sensor networks are also discussed in this section of the book. The details of both hypothetical sections are provided next.

After introducing the DSP architectures for embedded systems in Chapter 1, Chapter 2 provides co-synthesis techniques for embedded system design. Chapter 3 contains an overview related to non-intrusive dynamic application profiling and demonstrates how such runtime profiling can be effectively adapted to detect soft errors at the embedded application level. Chapter 4 investigates several software optimization techniques for embedded-processor systems. Power management has become one of the most important features in modern embedded systems. Chapter 5 discusses the power management framework that can be adapted to various real-time operating systems (RTOS) and explains a flexible power management support in RTOS based embedded systems. Chapter 6 delves into a detailed examination of the diverse benchmark strategies and will provide readers with a good understanding of where they need to apply their resources when analyzing embedded-processor performance.

The second part of the book starts with a discussion related to an unrelenting technology evolution to further develop the telecommunications business into the next decade. Chapter 7 takes into account the evolution of storage, processing (devices), sensors, displays, and autonomic systems and discusses how such an evolution is going to reshape markets and business models. Chapter 8 proposes an on-demand topology reconfiguration approach for mobile sensor networks aimed at enhancing the connectivity and performance. In Chapter 9, a network structure for delay-aware applications in wireless sensor networks is proposed where such structures are important for time-sensitive applications. Chapter 10 presents models for camera selection and handoff in a video network based on game theory. In Chapter 11, we deal with the creation and maintenance of mobile devices networks. Finally, Chapter 12 addresses the targeting of multiple mobile sensors in the context of ensemble forecasting.

Our book is useful for advanced graduate research work for the academicians and researchers. In addition, it will appeal to practicing engineers, senior undergraduates, and graduate students. This is an advanced book that can be used selectively in the courses for the final year of undergraduate study.

It deals with system level design techniques and covers both software and hardware aspects of embedded and networking systems. Tradeoff between system power and performance is addressed and co-design methodologies for real-time embedded systems are also covered in the book. The book finally addresses various types and structures of sensor networks covering mobile sensor networks and camera selection in video networks.

About the Editors

Gul N. Khan graduated in electrical engineering from the University of Engineering and Technology, Lahore in 1979. He received his MSc in computer engineering from Syracuse University in 1982. After working as research associate at Arizona State University, he joined Imperial College London and completed his PhD in 1989. Dr. Khan joined RMIT University, Melbourne in 1993. In 1997, Dr. Khan joined the computer engineering faculty at Nanyang Technical University, Singapore. He moved to Canada in 2000 and worked as associate professor of computer engineering at University of Saskatchewan before joining Ryerson University. Currently, he is a professor and program director of computer engineering at Ryerson University. His research interests include embedded systems, hardware/software codesign, MPSoC, NoC, fault-tolerant systems, high performance computing, machine vision, and multimedia systems. He has published more than eighty peer reviewed journal and conference papers. Dr. Khan also holds three US patents related to automatic endoscope and he has been the editor of three conference proceedings. Dr. Khan is currently serving as the associate editor of *International Journal Embedded and Real-Time Communication Systems*. He can be reached at gnkhan@ryerson.ca and his webpage URL is http://www.ee.ryerson.ca/~gnkhan.

Krzysztof (Kris) Iniewski is the managing director of research and development at Redlen Technologies Inc., a start-up company in Vancouver, Canada. Redlen's revolutionary production process for advanced semiconductor materials enables a new generation of more accurate, all-digital, radiation-based imaging solutions. Dr. Iniewski is also president of CMOS Emerging Technologies Research (www. cmosetr.com), an organization of high-tech events covering communications, microsystems, optoelectronics, and sensors. In his career, Dr. Iniewski has held numerous faculty and management positions at University of Toronto, University of Alberta, SFU, and PMC-Sierra Inc. He has published more than 100 research papers in international journals and conferences. He holds eighteen international patents granted in the United States, Canada, France, Germany, and Japan. He is a frequent invited speaker and has consulted for multiple organizations internationally. He has written and edited several books for IEEE Press, Wiley, CRC Press, McGraw-Hill, Artech House, Cambridge University Press, and Springer. His personal goal is to contribute to healthy living and sustainability through innovative engineering solutions. In his leisure time, Dr. Iniewski can be found hiking, sailing, skiing, or biking in beautiful British Columbia. He can be reached at kris.iniewski@gmail.com.

Contributors

Bir Bhanu
Center for Research in Intelligent
    Systems
University of California
Riverside, California

Jeff Caldwell
Embedded Microprocessor Benchmark
    Consortium
Mountain View, California

Chi-Tsun Cheng
The Hong Kong Polytechnic University
Hong Kong, China

Han-Lim Choi
Division of Aerospace Engineering
Korea Advanced Institute of Science
    and Technology
Daejeon, Republic of Korea

Muhamed Fauzi
Centre for High Performance
    Embedded Systems
Nanyang Technological University
Singapore

Chris Fournier
Embedded Microprocessor Benchmark
    Consortium
Mountain View, California

Gene A. Frantz
Department of Electrical and
    Computer Engineering
Rice University
Houston, Texas

Shay Gal-On
Embedded Microprocessor Benchmark
    Consortium
Mountain View, California

Mostafa Halas
Faculty of Engineering
Benha University
Benha, Egypt

Hicham Hatime
Electrical Engineering and Computer
    Science Department
Wichita State University
Wichita, Kansas

Jonathan P. How
Department of Aeronautics and
    Astronautics
Massachusetts Institute of Technology
Cambridge, Massachusetts

Mostafa E. A. Ibrahim
Institute of Telecommunications
Vienna University of Technology
Vienna, Austria

Siew-Kei Lam
Centre for High Performance
    Embedded Systems
Nanyang Technological University
Singapore

Francis C.M. Lau
The Hong Kong Polytechnic University
Hong Kong, China

Markus Levy
Embedded Microprocessor Benchmark
    Consortium
Mountain View, California

Yiming Li
Department of Electrical Engineering
University of California
Riverside, California

Ernesto López-Mellado
CINVESTAV Unidad Guadalajara
Zapopan, México

Roman Lysecky
Electrical and Computer
    Engineering Department
University of Arizona
Tucson, Arizona

Andres Mendez-Vazquez
CINVESTAV Unidad Guadalajara
Zapopan, México

Alexander Mintz
Embedded Microprocessor
    Benchmark Consortium
Mountain View, California

Kamesh Namuduri
Electrical Engineering Department
University of North Texas
Denton, Texas

J. Guadalupe Olascuaga-Cabrera
CINVESTAV Unidad Guadalajara
Zapopan, Mexico

Félix Ramos-Corchado
CINVESTAV Unidad Guadalajara
Zapopan, México

Markus Rupp
Faculty of Engineering
Benha University
Benha, Egypt

Roberto Saracco
Telecom Italia Future Centre
Venice, Italy

Karthik Shankar
Electrical and Computer Engineering
    Department
University of Arizona
Tucson, Arizona

Thambipillai Srikanthan
Centre for High Performance
    Embedded Systems
Nanyang Technological University
Singapore

Chi K. Tse
The Hong Kong Polytechnic University
Hong Kong, China

John Watkins
Electrical Engineering and Computer
    Science Department
Wichita State University
Wichita, Kansas

1 Evolution of Digital Signal Processing Architecture at Texas Instruments

Gene A. Frantz

CONTENTS

1.1 INTRODUCTION

I have watched the evolution of the digital signal processor (DSP) from an enviable position as an insider [1]. Two warnings should be noted before I begin a discussion of the history of DSP at Texas Instruments (TI). First, I am not a DSP architect. I am a user of architectures. What drives me is looking for the systems that can be designed with more capability at lower cost and ultimately achieve portability. As an aside, the U.S. Army's definition of *portable* when I was a soldier was "two privates can carry it." Fortunately, portable has come to mean "I don't even know I'm carrying it."

Second, I will relate the history of DSP from the view of my personal involvement. Many others will see this history differently and their stories will be as interesting as mine (assuming mine is interesting).

1

The story starts with the Speak & Spell™ learning aid, which has been described as the start of the era of the DSP. Then I will discuss the subsequent architectures introduced by TI. Finally, the vision of where we can go from here will be discussed.

1.2   SPEAK & SPELL DSP

The year was 1976 when Paul Breedlove proposed a novel educational product that could teach children how to spell. It was initially called the "Spelling Bee" and later introduced as the Speak & Spell learning aid [2,3]. It only had one minor technical issue that prevented it from being a tremendous success—it needed to be able to talk.

The architecture of the Speak & Spell DSP resulted from the collaboration of two individuals. One was Richard Wiggins, a new TI employee who was a speech research scientist. The second was a calculator integrated circuit architect named Larry Brantingham. Their job was to make the Speak & Spell product talk. The concept was to use linear predictive coding (LPC) with ten coefficients [4,5]. To achieve this, the team was required to use a very complex process known as 7 μM PMOS. Figure 1.1 shows a block diagram of the architecture and a die photo of the TMC0280 device.

Before jumping into the programmable DSP devices that TI introduced, here is a perspective that may help you while reading this chapter. It is material excerpted from a book about the history of TI's DSP devices [6]. The main feature of the book was the inclusion of an actual die from each generation of DSP. The book referred to the four dynasties of TI's product line. The next section discusses the background of the Speak & Spell and the development of the dynasties as detailed in Reference [6].

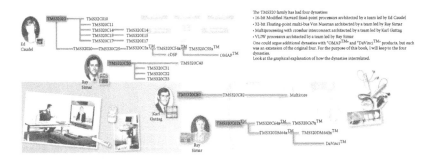

1.3   INTRODUCTION TO FIXED POINT DSPS

The Speak & Spell DSP was a fixed function device that performed only LPC-10 speech synthesis. It was obvious to several TI designers that although the TMC0280 was very successful, the next device needed to be more flexible. Specifically it needed to go beyond speech synthesis and perform speech analysis for vocoding, speech recognition, and speaker identification. Several design teams began architectural development of the next device. TI's Educational Products Division wanted a follow-on device to create products using speech encoding to further enhance its learning aid

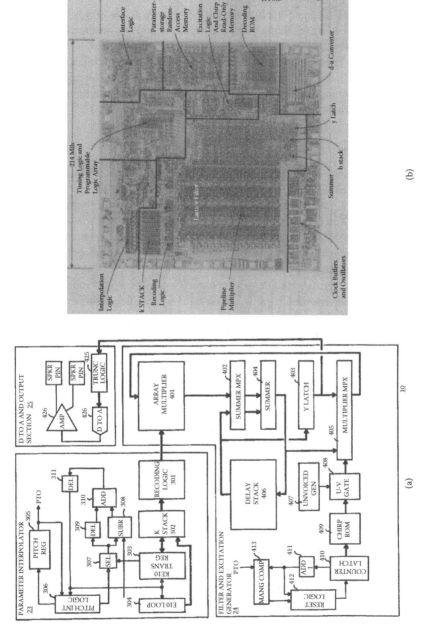

FIGURE 1.1 (a) Block diagram of TMC0280 Speak & Spell speech synthesizer from U.S. Patent 4,209,836. (b) Die photo of device.

product line. Another team worked on a more general purpose architecture that could be programmed for speech algorithms and perhaps telecommunications algorithms.

After great numbers of interactions among device architects, the final decision was to focus on the architecture that showed the greatest flexibility. The choice was to modify a Harvard [21] architecture having 32-bit instruction words and 16-bit data words. The Harvard architecture was chosen because (1) it was the architecture chosen for the TMS1000 microcontroller product line TI already had in production; and (2) it could use the instruction and data buses after minor modifications to act as data buses that allowed simultaneous feeding of both the multiplier and multiplicand to the hardware multiplier. TI cited the "Modified Harvard" architecture in its market communications [22].

1.3.1 TMS32010

The first programmable DSP known as the TMS32010 (Figure 1.2) was introduced at the International Solid-State Circuits Conference (ISSCC) in 1982 [7]. Figure 1.3 makes it easy to describe the architecture by dividing the die into four quadrants: CPU, RAM, ROM, and multiplier. Later in the chapter we will show die photos of later generation DSPs in which the multipliers are not easily detectable because of size reductions resulting from greater density achieved by advanced integrated circuit fabrication processes.

One heated architectural arguments concerned the need for interrupts. The initial design did not have an interrupt based on the very simple reasoning that real-time systems, by definition, could not be interrupted. The argument was settled by including an interrupt that could be polled by the CPU when it was not involved in a real-time process.

Until then, all TI's processors had two versions—one for development and one for production. The emulator was designed to allow system designers to design and debug their products. The second (production) device was used in the manufacturing of their end products. The TMS32010 was to be one of the first self-emulation devices. That meant the 2.5 K on-chip ROM could be replaced by external SRAM until the code was mature enough to create a ROM at mask level to program the on-chip memory. When we learned that many of our customers used the emulation mode to turn a ROM-based device into a RAM-based device, we realized we were more brilliant than we thought.

One last point must be made about the size of the on-chip RAM—it was only 144 words. The two reasons for choosing that RAM size were (1) we ran out of die size to increase, and (2) the chosen size was large enough for a 64-point FFT and still allowed sixteen words of scratch pad memory for the programmer to use. The CMOS versions of the TMS32010 were expanded to 256 words of on-board RAM.

Figure 1.2 is a block diagram of the TMS32010 architecture [7]. Table 1.1 lists the various generations and a few design details of the fixed point DSPs marketed by TI. The list in Table 1.1 is by no means complete. TI continues production of many of the devices in these product families. Further, the TMS320C25 was the first in a family of digital signal controllers designated the C24x, C27x, and C28x. They are now known as microcontrollers. The TMS320C50 device allowed TI to enter the digital

TMS 32010 FUNCTIONAL DIAGRAM

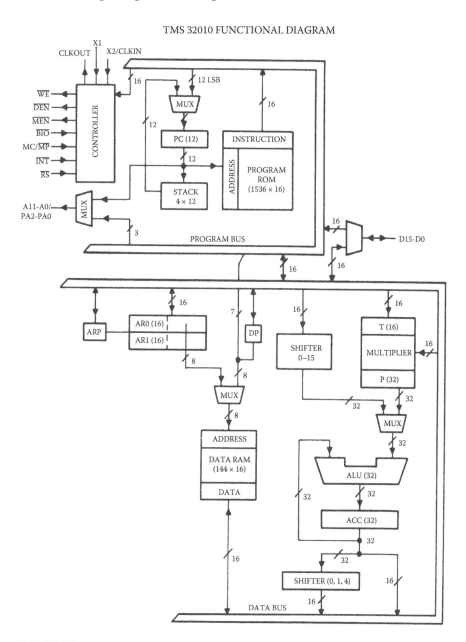

FIGURE 1.2   TMS32010 block diagram.

cell phone market and began a family of phone architectures (C54x, C55x, and OMAP). I'll discuss the OMAP architecture in Section 6.1 covering multi-processing.

The birth of the digital cellular phone market finally drove TI to focus on power dissipation as a performance metric. The focus actually started in the DSP product team a few years earlier.

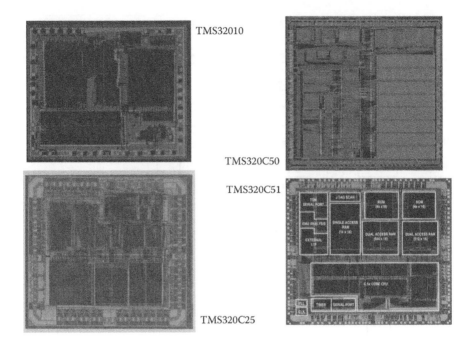

FIGURE 1.3   Die photos of TMS32010, TMS320C25, TMS320C50, and TMS320C51.

Sometime in the late 1980s, a customer called us about our first-generation devices. The customer used them in conjunction with hearing aids and requested TI to ship the 3V version of the device. In a typical vendor response, we informed them that we had only 5V devices that did not work at 3V. The customer, however, was certain that we had 3V devices because all the products purchased from TI ran without failure. After listening to the request, TI application engineers conducted a study and realized that, in fact, the 5V devices ran at 3V. A data sheet was quickly created and after convincing the business team, TI offered a 3V device.

When the digital cellular phone market began to accelerate, another customer left TI a simple message: "If you can't lower the power dissipation on your devices, we'll find another vendor." By this time listening to customers became a standard practice at TI (Figure 1.3).

Allow me to make this discussion more personal since I was the application manager for the device and a member of the start-up team for the digital cellular business unit. The customer request for better power dissipation encouraged me to look back at our DSP history and better understand power dissipation. To my surprise, the plot of power dissipation per millions of multiply accumulates per second (MMACS) revealed an amazing trend of 50% reduction every eighteen months. This graph was used within and outside of TI to drive power dissipation down as an active design issue rather than as an incidental result. The drastic reduction was called "Gene's Law" [1] as a humorous way to get the attention of the designers and also to

TABLE 1.1
List of 16-Bit Fixed Point Product Families

Family	Technology	Die Size	Transistor	Performance	Introduction Year	Architect
TMS3201 [7]	3u NMOS	43.81 mm^2	58 K	5 MIPS	1982	Ed Caudel
TMS32020 [8]	2.4u NMOS	74 mm^2	80 K	5 MIPS	1985	Surendar Magar
TMS320C25 [9, 10]	1.8u CMOS	8.9 × 8.9 mm	100 K	10 MIPS	1986	Takashi Takamizawa
TMS320C50 [11]	0.5u CMOS	14 × 14 mm	1.2 M	20 MIPS	1992	Peter Ehlig

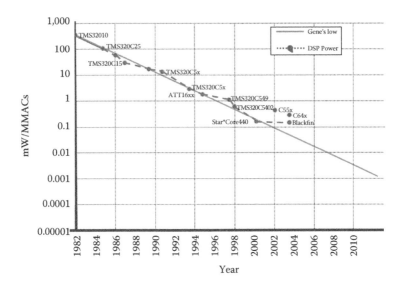

FIGURE 1.4   Gene's Law showing relationship of DSP performance and power dissipation over time.

signify the importance of power dissipation as a new way to measure performance. Figure 1.4 shows Gene's Law as initially constructed.

Before leaving this section about fixed point processor architecture, an interesting story that lacks proof is still worth telling. When the Iron Curtain came down in the mid-1980s, we learned that the TMS320C20 was the standard DSP of the Eastern European world. We assumed the original TMS32020 had been reverse engineered and put into CMOS devices. We didn't know whether to be proud of or angry about the unauthorized use of TI's product.

1.3.2   FLOATING POINT

While the DSP technical community was convinced that signal processors should be fixed point types programmed only in assembly code, TI's architect, Ray Simar, began development of the TMS320C30. Integrated circuit (IC) technology had advanced far enough to make both 32-bit floating point and high level language programming possible.

The modified Harvard architecture was replaced with a multi-bus Von Neumann architecture and progressed to a 32-bit data word. With the data word and instruction word of the same size, both could be stored in any of the memory blocks in the device or external to the device. The multiple bus architecture allowed multiple parallel accesses so that the multiplier could be fed in parallel to fetching instruction words. The internal buses [12–14] were:

Program (32-bit data, 24-bit address)
Data with two address buses (32-bit data, 24-bit address)

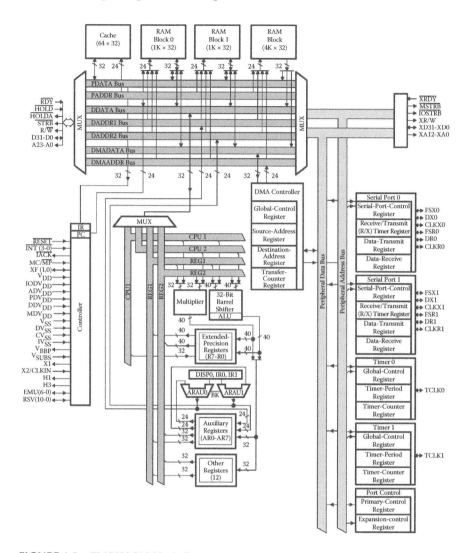

FIGURE 1.5 TMS320C30 block diagram.

DMA (32-bit data, 24-bit address)
Peripheral (32-bit data, 13-bit address)

The TMS320C30 (Figure 1.5) had two external ports, one with 24 bits of address reach and one with 13 bits of address reach. It was introduced at the International Conference on Acoustics, Speech, and Signal Processing (ICASSP) in 1989, about three months ahead of schedule, a result of a surprise first pass success in wafer fabrication. It not only caught us off guard, but many of our customers were surprised because TI had not built the prototype systems the C30 plugged into.

The use of then current IC technology node (1-micron CMOS) allowed the inclusion of more on-board memory into the device. More memory allowed processing of larger datasets and employment of more sophisticated algorithms. Specifically, the C30 contained two dual-access 1-K RAM blocks, one 4-K dual-access ROM block plus 64 words of cache memory.

About a year into the production life of the TMS320C30, it became obvious that most system designers used multiple C30s in their designs. It seemed that we finally introduced a DSP that changed the thought processes of the designers who used it. Before C30 devices became common, the designers asked, "Can I squeeze my algorithm into the given space?" The C30 changed the question to, "When I divide my problem by the performance of a C30, is the number of devices small enough for me to handle?" The DSP design made a significant impact. After considering some reasonable requests from customers, Ray architected the TMS320C40.

1.4 DSP ARRAY PROCESSING

The TMS320C40 displayed several improvements over the TMS320C30 that made it attractive for multi-processing systems [15]:

- Six parallel communications ports
- Two external buses, one for sharing and one private
- Fully compliant IEEE floating point

The six parallel 8-bit communications ports allowed many interesting multi-processing configurations, the most sophisticated of which was the formation of a cubic interconnect. The issue was simple: not all users could know how to use the parallel ports in their systems reliably. It was even thought that it was impossible to make them work in a system design. In fact, the belief then was that good designers made systems work where average and poor designers could not.

The two external buses allowed each C40 in a multi-processing design have its local memory for its exclusive use while sharing a common memory with other processors in the system. Obviously the shared memory block allowed a way for one C40 to communicate with another. The address reach of each of these two memory spaces was 2 gigawords (32 bits) or a total of 4 gigawords of total memory reach for the C40 processor.

It was a surprise that the lack of IEEE floating point compatibility created a need for system designers who used the C30 so this feature was added as a C40 data format (Figure 1.6). In addition, a 40-bit extended precision floating point format was included.

After the TMS320C40 was introduced, TI turned its architectural focus back to fixed point design with the TMS320C62x. However, in the background, a next generation floating point device called the TMS320C7x was developed quietly and reached the commercial market as the TMS320C67x.

The reason for turning back to fixed point design was the idea that fixed point represented "real" DSP opportunities for performance and cost reasons. More than a decade passed before floating point again became an architectural focus of TI.

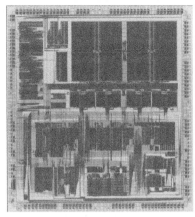
TMS320C30

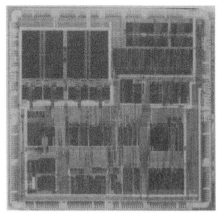
TMS320C40

FIGURE 1.6 Die photos of TMS320C30 and TMS320C40.

TABLE 1.2
List of 32-Bit Floating Point Product Families

Family	Technology	Die Size	Transistor	Performance	Introduction Year	Architect
TMS320C30	1 µ CMOS	12 × 13 mm	250 K	32 MFLOPS	1989	Ray Simar
TMS320C40	0.72 µ CMOS	11 × 11 mm	500 K	40 MFLOPS	1991	Ray Simar
TMS320C67x	0.18 µ CMOS	13 × 15 mm	1.2 M	200 MHz	1997	Ray Simar

In fact, TI's high performance, multi-core processors known as the TMS320C66x family are all floating and fixed point processor cores. These devices will be covered in the section on multi-processing DSPs.

Note that the last device listed in Table 1.2 is a very large instruction word (VLIW) architecture. It will be covered in Section 1.5.

1.5 VERY LARGE INSTRUCTION WORD (VLIW) DEVICES

Before spending time talking about multi-processing architectures, it is worthwhile to introduce the TMS320C6x family of VLIW devices [25]. Ray Simar finished the architecture of the TMS320C40 in the early 1990s and began what would be the next generation of DSP architecture. In his early research, it became clear that an interesting architectural concept of the past known as VLIW became practical with the advancement of IC technology [26]. The first device introduced in this family was the TMS320C6201. Some of its architectural aspects [16] were:

- Two data paths with (almost) identical functional blocks
- Four functional units per data path
  - .L unit:
    - 32/40 bit arithmetic and compare operations
    - 32-bit logic operations
  - .S unit
    - 32-bit arithmetic operations
    - 32/40 bit shift and 32-bit bit-field operations
    - Branches
    - Constant generation
    - Register transfers to and from control register file (.S2 only)
  - .M unit
    - 16 × 16 bit multiply operations
  - .D unit
    - 32-bit add, subtract, linear, and circular address calculations
    - Loads and stores with 5-bit constant offset or 15-bit constant offset (.D2 only)
- 32-bit instruction word for each functional unit

Figure 1.7 shows the overall block diagram of the TMS320C6201, a die photo of the device, and a block diagram of the functional units.

1.6 MULTI-PROCESSING

In this chapter, the *multi-processing* term will be used rather than multi-core since the drive for more performance has led DSP architecture down multiple paths. Only one path can be described as multi-core. Here is a quick summary of multi-processing architectural paths:

- Custom solution
- Heterogeneous multi-processing
- Homogeneous multi-processing

1.6.1 CUSTOM SOLUTIONS

This was the first architectural path taken at TI (see Section 1.2). Two examples are worth discussing. The first was an offering to the industry in the early 1990s of a TI product designated Custom DSP (cDSP). The second was a specific device introduced for the television market before the advent of digital TV. It was called the Serial Video Processor (SVP, Part TMC57100) [17]. Figure 1.8a is a die photo of the SVP.

The cDSP was intended to fulfill many of TI's high volume vertical market opportunities. It allowed system designers to integrate onto one circuit a fixed point DSP, a block of gate array logic, and simple analog interfaces (A/D or D/A). Each device

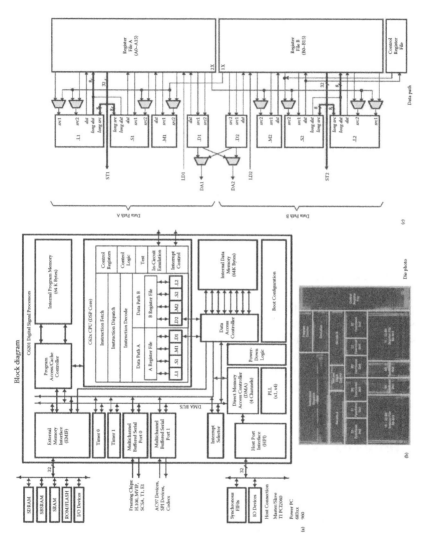

FIGURE 1.7 (a) Block diagram of TMS320C6201. (b) Die photo of device. (c) Block diagram of functional units.

TMS57100   OMAP5910

 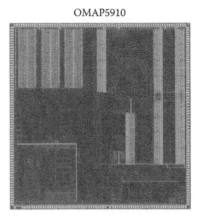

FIGURE 1.8   (a) Die photo of TMS57100. (b) Die photo of OMAP5910.

was unique so the gate array could be of any size including multiple sub-arrays. One of the "Aha!" moments was the realization that a DSP algorithm executed in custom logic gave the highest performance and consumed the lowest power at the lowest cost.

Around that time, a team of engineers in the DRAM business unit developed a multi-core processor to allow an NTSC TV signal to be processed digitally. It was called the SVP. The architecture was simple. It consisted of a thousand 1-bit DSPs installed on one integrated circuit. After each horizontal line of video was converted into digital pixels, they were fed in parallel to the DSPs. The design had a single instruction, multiple data (SIMD) architecture. Each processor processed a different pixel, but in exactly the same way as the other processors. After the pixels were processed, they were converted back into analog form for output. The SVP was well received in the TV industry and millions were sold for high end sets.

An observation at this point about multi-processor architectures is that if a device is designed for specific end equipment, many issues are minimized or even eliminated.

1.6.2   Heterogeneous Multi-Processing

Heterogeneous multi-processing is best described as a set of processing elements that work together to optimally solve a significant (unsolvable) problem or create an opportunity by doing something that has never been done before. As stated earlier, the optimal solution to any problem using IC technology is a custom design. It can yield the highest performance, lowest cost, and lowest power dissipation.

In a system architecture, portions of the system will be well defined, some parts will have some variability, and some will provide infinite opportunity for improvement. Each type must be handled differently. For example, in a communications system, several algorithms may best fit into an accelerator (i.e., a fixed function block). Examples of such algorithms might be Viterbi encoding and decoding and Reed Solomon encoding and decoding. The algorithms used in a video or audio compression system may be a fast Fourier transform or Huffman coding. For many years I called these accelerators "semi-programmable" processors.

OMAP™ (Open Media Applications Platform) was the first heterogeneous multi-processing family of devices that TI developed. It was initially designed only for the digital cell phone market, but later became a perfect fit for smart phones. It has since expanded into many other embedded markets. All of devices in the OMAP family are ARM-based [19]. Table 1.3 gives a quick overview of the generations of the OMAP family of system processors and Figure 1.8b is a die photo from the first generation.

1.6.3   HOMOGENEOUS MULTI-PROCESSING

In one respect, the homogeneous multi-core is a simple concept. Most technologists use the *multi-core* term to describe multiple placements of one type of core on the same silicon substrate.

The first multi-core DSP introduced by TI was the TMS320C80 [24]. It consisted of four signal processors and a host processor all connected through a cross bar switch. Figure 1.9 shows a high level block diagram and a die photo.

1.7   WHAT NOW?

1.7.1   FIXED POINT DSPS

- The TMS320C2x family remains a popular solution for motor control along with other control applications. The newest in this family is the TMS320C28xx that has added floating point to its list of features along with flash memory and a wide variety of analog and digital interfaces.
- The TMS320C5x family remains a popular solution for audio band signal processing applications. Its latest, the TMS320C55x, is also designed for low power operation.
- In addition to these two families, a third family called the MSP430 is a popular solution to many ultra-low-powered applications such as medical implants.

1.7.2   FLOATING POINT DSPS

- The TMS320C30 and TMS320C40 are still in production and in some cases still used for new designs.
- The TMS320C67x is still a popular floating point DSP used in many high end audio applications.
- The latest version of VLIW architectures, the TMS320C66x, has both fixed and floating point capabilities.

1.7.3   MULTI-PROCESSING DSPS

- The TMS320C80 and its smaller version, the TMS320C82, are still in production but are not used for new designs.
- The TMS57100 and cDSP capability are no longer offered to the market.

TABLE 1.3
OMAP Summary by Generation

Generation	Introduction Year	Processor	ARM/Clock	DSP/Clock	Accelerator	Display	Camera
1	2002	130 nM	925/162 MHz	C55/162 MHz	D	176 × 144	0.3 M pxl
2	2006	90 nM	1136/330 Mhz	C64/220 MHz	V, G, D	640 × 480	1 × 3
3	2008	65 nM	A8/600 MHz	C64/300 MHz	V, G, D	1280 × 800	2 × 8
4	2010	45 nM	2 A9s/1.2 GHz	C64/500 MHz	V, G, D	1920 × 1200	2 × 16
5	2012	28 nM	2 A15s/1.8 GHz	C64/532 MHz	V, G, D	2048 × 1536	5 × 24

V = Video accelerator; G = Graphics accelerator; D = Display accelerator.

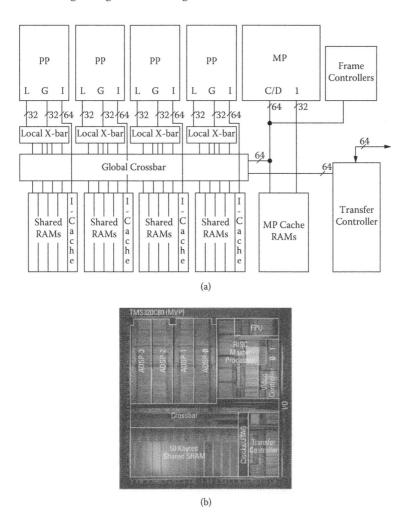

FIGURE 1.9   TMS320C80. (a) High level block diagram. (b) Die photo.

- OMAP is still actively being designed into new applications. The fifth generation is ready for design.
- The TMS320C66x family [18] of processors includes several multi-core versions. They can be found in many communications infrastructure systems and in other devices that require high performance with low power dissipation.

1.8   CONCLUSION

I have been quoted saying that "DSP is dead as a product, but all future processors will be enabled by DSP." I made that statement many times—not to predict the demise of the DSP, but to remind all of us that we are just beginning the path of taking real world signals and processing them to do things that have never been done before. I also quote a professor friend, Dr. Alan Oppenheim, when I asked him whether DSP was really finished. His response was, "Gene, there will always be interesting signals that we will want to process. Therefore there will always be a need for signal processors."

The future is bright for signal processing and for microprocessors that can process real world signals. Performance and power dissipation will continue to drive new architectures. The DSP's capability has invaded the world of embedded processors. It is also likely that the DSP will re-invade the world of analog processors. In both cases, DSPs will be bounded by Moore's law [20] and Amdahl's law [23]. In summary:

"It will be Moore's law that determines how many transistors can be integrated onto one integrated circuit, but it will be Amdahl's law that determines how the transistors are used."

I hope this history of the contributions Texas Instruments made to the digital signal processor market has been informative. It was a pleasure to have been part of it. Remember one thing: the era of digital signal processing is not over; it is taking new directions. We live in exciting times. And we will continue to search for new and interesting signals to process.

REFERENCES

1. Frantz, G.A. 2012. A short history of the digital signal processor. *IEEE Solid State Circuits*, 4, Spring.
2. Frantz, G. and Wiggins, R. 1989. The development of "solid state speech" technology at Texas Instruments. *ASSP Newsletter*, 53, 34f.
3. Frantz, G.A. and Wiggins, R.W. 1982. Design case history: Speak & Spell learns to talk. *IEEE Spectrum*, February, 45–49.
4. Wiggins, R. and Brantingham, G.L. 1978. Three-chip system synthesizes human speech. *Electronics Magazine*, August, 109–114.
5. Brantingham, L. 1979. Single-chip speech synthesizer and companion 131 K bit ROM. *IEEE Transactions on Consumer Electronics*, 25, 193–197.
6. Frantz, G.A. 2011. *TI's DSP History in Die Form*, Texas Instruments. http://cnx.org/content/m44975/latest/
7. Magar, S.S., Caudel, E.R., and Leigh, A.W. 1982. A microcomputer with digital signal processing capability. In *Proceedings of International Solid-State Circuits Conference*, pp. 31–33 and 284–285.
8. Magar, S., Essig, D., Caudel, E. et al. 1985. An NMOS digital signal processor with multi-processing capability. *ISSCC Digest*, February, 90–91.
9. Abiko, S., Hashizume, M., Matsushita, Y. et al. 1986. Architecture and applications of a 100-nS CMOS VLIW digital signal processor. In *Proceedings of International Conference on Acoustics, Speech, and Signal Processing*, pp. 393–396.
10. Frantz, G.A., Lin, K.S., Reimer, J.B. et al., Bradley, J, The Texas Instruments TMS320C25 digital signal microcontroller. *IEEE Micro*, 10–28.
11. Lin, K.S., Frantz, G.A., and Lai, W.M. 1989. Texas Instruments' newest generation fixed point DSP: the TMS320C5x. International Symposium on Computer Architecture and Digital Signal Processing, October.

12. Simar, R., Leigh, A., Koeppen, P. et al. 1987. A 40 MFLOPs digital signal processor: the first supercomputer on a chip. In *Proceedings of Internal Conference on Acoustics, Speech, and Signal Processing,* pp. 535–538.

13. Texas Instruments. 1991 (revised 2004). SMJ320C30 Digital Signal Processor Data Sheet, SGUS014.

14. Lin, K.S., Frantz, G. A., and Simar, L.R. 1987. The TMS320 family of digital signal processors. *Proceedings of IEEE*, September.

15. Jain, Y. 1994. Parallel Processing with the TMS320C40 Parallel Digital Signal Processor. Texas Instruments Application Report SPRA053, February.

16. Texas Instruments. 1997. TMS320C6201 Fixed Point Digital Signal Processor Data Sheet, SPRS051.

17. Childers, J., Reinecke, P., Miyaguchi, H. et al. 1990. SVP: serial video processor. In *Proceedings of IEEE Custom Integrated Circuits Conference*, pp. 17.3/1–17.3/4.

18. Texas Instruments. 2012. TMS320C6678 Multi-Core Fixed and Floating-Point Digital Signal Processor Data Manual. SPRS691C.

19. http://en.wikipedia.org/wiki/OMAP

20. Moore, G.E. 1985. Cramming more components onto integrated circuits. *Electronics,* 38, April 19.

21. http://en.wikipedia.org/wiki/Harvard_architecture

22. Texas Instruments. 1985. *TMS32010 Users Guide.*

23. Amdahl, G.M. 1967. Validity of the single-processor approach to achieving large scale computing capabilities. In *Proceedings of AFIPS Conference*, pp. 438–485.

24. Gove, R.J. 1994. The multimedia video processor (MVP): a chip architecture for advanced DSP applications. *IEEE Digital Signal Processing Workshop*, pp. 27–30.

25. Fisher, J.A. 1983. *Very Long Instruction Word Architectures.* New Haven: Yale University.

26. Simar, R. and Tatge, R. 2009. How TI adopted VLIW in digital signal processor. *IEEE Solid State Circuits*, 1, 10–14.

2 Co-Synthesis of Real-Time Embedded Systems

Gul N. Khan

CONTENTS

2.1 INTRODUCTION

Most modern electronic devices are embedded systems. A luxury car contains more than 100 embedded processors that are responsible for its antilock braking system (ABS), electronic data system (EDS), traction control system (TSC), electronic stability control), security, alarms, and related activities [1]. Microwave ovens, dishwashers, and washing machines used in households, smart phones we use for communications and entertainment, and multimedia devices are based on embedded systems [2].

Some of these systems along with high performance embedded systems used in aerospace, autonomous guidance, nuclear, and industrial control applications are multitask embedded systems that support applications involving multiple independent executable tasks. Each task requires multiple function modules that are presented as processes in this chapter.

A task can involve multiple processes of the same functionality and different tasks may require processes of the same functionality. Designing a high performance embedded computer system, especially systems with real-time constraints, is a challenge. Life-supporting medical equipment, nuclear power plants, and flight control systems are examples of such embedded systems. Many of these systems utilize task levels and explicit process level deadlines. A designer must ensure that real-time deadlines are met either by faster processors or dedicated hardware. However, designers must also design and develop affordable embedded systems.

Recent implementations of embedded systems consist of general purpose processors (software processing elements) and dedicated hardware blocks (hardware processing elements) such as application-specific integrated circuit (ASICs) and/or field programmable gate arrays (FPGAs) [3]. Manual design space exploration for embedded systems is time consuming and expensive because it requires design and testing of numerous prototype embedded systems.

Although embedded computer systems containing multiple processing elements (PEs) are gaining popularity, design automation of these systems poses a number of challenges. We propose a novel design automation method that can produce optimized partitioned (hardware and software) solutions for large-scale, multiple-processing-element, multitask embedded systems. Initial design automation focused on low level stages of the design process and later moved toward the automation of high level (architectural) design processes. Problem definition for high level design is far less clear than it is for low level system design.

This chapter is organized in five sections. Section 2.2 provides an overview of the hardware–software co-synthesis problem for embedded computer systems, optimization methods for scheduling to meet real-time constraints, and co-synthesis and co-design techniques proposed by various research groups. In Section 2.3, we describe phases of the proposed co-synthesis method for real-time embedded systems. These include various techniques for the initialization of PEs, communication resource allocation, scheduling of processes on allocated PEs, and evaluation of newly generated solutions.

Section 2.4 presents and discusses the experimental results for our co-synthesis methodology. The proposed methodology is employed to generate solutions for various embedded application task graphs from the literature and some well-known benchmarks. The embedded system solutions for these applications are also evaluated and the results compared with previous work. We summarize the contributions and propose directions for future work in Section 2.5.

2.2   CO-SYNTHESIS AND REAL-TIME CONSTRAINTS

Hardware–software co-synthesis automates the process of design space exploration for embedded computer systems requiring hardware or other PEs in addition to one traditional CPU. The problem of mapping functionalities onto one of the available dedicated hardware processing elements (hardware modules) or implement them as processes on one of the available software processing elements (traditional CPUs) can be viewed as an optimization problem. The goal is to optimize implementation cost and performance while meeting real-time constraints of the system. Four main steps are carried out by a CAD (co-synthesis) tool; namely, allocation, assignment, scheduling and evaluation:

- Allocation selects various types of PEs and communication links that can be used in the system.
- Assignment (also known as binding or mapping) determines the mapping of processes to PEs and communicates events to links.
- Scheduling sets the order in which processes will be executed on the assigned PEs.
- Evaluation determines the cost, performance, and amount of deadline violation of the embedded system solutions.

To achieve optimal solutions, it is important to have feedback from one step of co-synthesis to the others as these steps influence each other.

2.2.1   PRELIMINARIES

The proposed co-synthesis methodology is intended to design multitask embedded systems to support applications involving multiple independent executable tasks. For our co-synthesis method, the functional system requirements and constraints are represented by a task set in which each task consists of multiple function modules. Fixed point bit allocation (FPBA) and convolutional encoder (CE) are examples of function modules for telecom applications.

Functional modules for a system are presented as processes in this chapter. A task can utilize multiple processes of the same functionality and different tasks can involve processes of the same functionality. We assume the processes are coarse grained—they are sufficiently complicated to require numerous CPU instructions.

A directed acyclic graph (DAG) represents a task of a system. The nodes of a DAG represent the processes of a task. Figure 2.1 depicts a simple example of a system with two tasks and related processes. First task has four processes; the second has

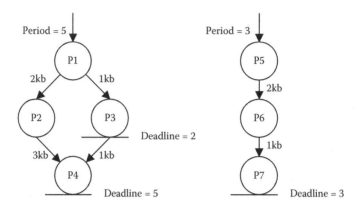

FIGURE 2.1   Task list representing system specification.

three processes. The volume of data transferred between a source and a destination node is associated with the edge (arc) between the two nodes.

A dependent process may start executing after all its data dependencies have been satisfied. Process P4 of the first task may begin execution after both processes P2 and P3 completed their execution and required data to process P4 (3 kbits from process P2 and 1 kbit from process P3) have been transferred. The *period* of a task is the time between one copy of task-ready time and the next consecutive copy of the same task-ready time. A system with multiple tasks may have different periods for different tasks.

The *hyperperiod* of a task set is the least common multiple of the periods of the tasks. A node without incoming edges is called a *source* node. A node with no outgoing edge is called a *sink* node. The deadline of a process (or task) is the time by which the process (or task) must complete its execution. A deadline can be soft or hard, depending on the real-time application. When one or more multiple processes violate the hard deadline, the solution will be invalid. A process can miss its soft deadline or meeting a soft deadline may not be mandatory.

A PE for executing processes may be a CPU, ASIC, FPGA or another type. A PE library is available as input to the CAD tool for co-synthesis of embedded systems. The PE library describes some characteristics of the PEs such as price, power consumption, etc. It also contains information about the relationships between processes and PEs such as worst-case execution time for each process–PE pair. For a specific embedded system architecture, the dependent processes may be assigned to different PEs. The edge connecting these processes must be assigned to a communication resource. If no communication resource connects the PEs, the architecture is invalid.

The main attributes of a communication resource are the communication controller cost, transmission time per bit, number of contacts, and power consumption during operation. A communication resource can connect multiple PEs that are equal or fewer than the number of contacts of the resource. A point-to-point communication resource has two contacts and a bus can have more than two contacts. The worst-case communication time for an edge can be calculated from various parameters such as the amount of data to be transferred for the edge and transmission time per bit for the communication resource.

2.2.2   OPTIMIZATION TECHNIQUES FOR HARDWARE–SOFTWARE CO-SYNTHESIS

Allocation, assignment, and scheduling steps of co-synthesis are known to be NP-complete problems [4,5]. A guaranteed optimal solution to an NP-complete problem usually takes a large amount of time that grows exponentially with the problem size. Heuristic algorithms have been used to solve co-synthesis problems for large-scale embedded computer systems. Optimization algorithms attempt to minimize system costs that depend on optimization parameters of the system.

The set of solutions around an embedded system solution is known as the solution neighborhood. The solutions can be reached in a single discrete step by the optimization method. A local minima is a solution that has the lowest optimized cost in the neighborhood. A global minima is a solution that has the lowest optimized cost in the whole problem space. In general, optimization algorithms try to avoid getting stuck in local minima and attempt to find global minima.

Some of the heuristic algorithms work iteratively to improve solutions. Greedy algorithms are among the simplest and most commonly used heuristic approaches. They tend to be trapped in local minima or local maxima. In this section, we discuss popular optimization techniques that can be used to solve the scheduling, assignment, and allocation problems.

Integer linear programming (ILP) is one of the earlier optimization techniques used for co-synthesis. To formulate an optimization problem for ILP, the objectives and constraints are defined as linear equations. An advantage of ILP formulation is that it provides an exact solution to a scheduling problem. However, it is computationally complex and fails to provide solutions for large-scale embedded systems.

Dynamic programming is another well known approach for finding optimal solutions. It decomposes a problem into a sequence of stages and the optimal solution to each stage must belong to the optimal solution of the original problem. The success of this approach depends on how efficiently a problem is divided. Like ILP, dynamic programming becomes computationally expensive for large-scale systems. Therefore, heuristic approaches have been widely researched to solve the process assignment and scheduling problems for co-synthesis. These heuristic approaches provide good quality solutions but cannot guarantee optimality.

Among the heuristic scheduling algorithms, list scheduling is widely suggested in the literature [6]. A priority list of the processes (based on some heuristic urgency measure) is used to select among the processes. The priority list can be modified to support timing constraints by reflecting the proximity of an unscheduled process to a deadline. This technique makes it possible to meet the real-time constraints. However, the heuristic nature of list scheduling does not ensure the finding of an optimal solution.

List scheduling can be applied to find a solution for minimum resources with latency constraint problems. At the beginning, least numbers of processors are allocated and additional processors can be allocated to meet latency constraints. The processes are prioritized through a slack-based method. Process *slack* is based on the difference between the latest possible start time of a process and the index of the schedule step under consideration. The low computational complexity of list scheduling makes it attractive for large-scale systems.

For process assignment and scheduling problems, heuristics techniques such as simulated annealing (SA), tabu search (TS), and genetic algorithms (GAs) are widely used. SA algorithms are iterative improvement techniques; greediness increases during their execution [7]. During the execution of an SA technique, the acceptance probability P of a modified solution may be represented by Equation (2.1).

$$P = \frac{1}{\left(1 + e^{\frac{N-P}{T}}\right)} \tag{2.1}$$

where T is the global temperature parameter; P is the cost of the old solution; and N is the cost of the modified solution.

The temperature parameter begins at a high value (e.g., infinity) and decreases as the system stabilizes. At the start of an SA algorithm run, the changes that increase the cost of a solution are selected with the same probability as the changes that decrease the cost. In this way, the algorithm can escape the local minima, but this failure to distinguish solutions based on cost does not help reach the goal of lower optimized cost. Toward the end of a run, the SA algorithm degrades to a greedy iterative technique.

TS methods are powerful heuristic techniques suitable for efficient hardware software co-synthesis [8]. TS uses a local search method to escape local minima and maintains a list that tracks all the recently visited solutions. The iterative improvement procedure prohibits revisiting of the solutions on the list to be revisited during the next iteration and thus tries to avoid getting trapped in one neighborhood of the solution space. After a specified number of iterations, the search is restarted from initial system state to expand the search of the design space.

GAs maintain a set of solutions and during their execution, solutions evolve in each generation. Previous solutions are improved by randomized local changes performed by genetic operators and exchanges of information between solutions. The lowest quality solutions of a generation are then removed from the set by a ranking and selection process [9]. A chromosome in a GA typically refers to a candidate solution to a problem.

All changes to chromosomes are made with two operators known as the *mutation* and *crossover*. *Mutation* operation randomly picks a location in the solution array and changes the entry of the location with a new value. *Crossover* randomly picks two solutions and their locations and swaps portions of solutions between these two locations.

Figure 2.2 shows an example of crossover. The randomly chosen positions are 1 and 3. After crossover portions of solutions between positions 1 and 3 are swapped, and a new pair of solutions evolves. The GA is capable of handling difficult problems composed of multiple NP-hard problems, even if each problem involves huge solution spaces and can share information between solutions. However, GA-based techniques are difficult to design and implement.

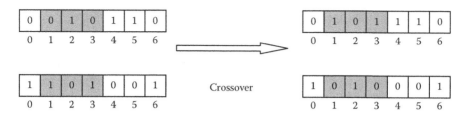

FIGURE 2.2 Crossover for genetic algorithm.

2.2.3 Hardware–Software Co-Synthesis Approaches

A lot of research has been done in the area of hardware–software co-synthesis. Some related research works on hardware–software (HW-SW) co-synthesis with real-time constraints have covered multimode–multitask embedded systems. Unfortunately, most of the techniques suffer from some problems. They are computationally intensive or do not account for real-time constraints, process level deadlines, communication link scheduling, and other issues. Prakash and Parker used ILP to solve the HW-SW partitioning problem [10]. It is possible to find the best optimal solution for embedded systems with a limited number of processes.

Gupta and Micheli presented one of the earlier approaches to optimization for hardware–software co-synthesis [11]. The authors proposed the idea of optimized HW-SW partitioning to meet the constraints imposed by the designer. They suggested examining a large number of solutions to find one that minimized this cost function.

Ernst et al. developed a co-synthesis tool known as COSYMA in which the system specification is represented in a superset of C language [12]. A simplified assumption of only one software processing element and one hardware module was made. The authors used SA for HW-SW partitioning. The method starts with infeasible solutions that violate the timing constraints, and then improves the solutions by migrating functionality from software to hardware. A cost function is used to control the partitioning process.

Hankel and Ernst also proposed a co-synthesis method that dynamically determines partitioning granularity to achieve better partitioning results [13]. They assumed that software and hardware parts executed in mutual exclusion and the partitioning process was integrated into COSYMA.

Liu and Wong integrated partitioning with scheduling for their iterative improvement co-synthesis algorithm [14]. They made a simple assumption of a fixed number of PEs for a system with two software PEs and n hardware PEs. The algorithm starts with allocating all the processes to the software PEs. Then appropriate processes are migrated to hardware based on feedback from the scheduler to minimize the completion time and resource cost.

Mooney and Micheli presented a tool that performs real-time analysis and priority assignments [15]. To meet the hard real-time constraints, they used dynamic programming to assign static priorities to the processes. They proposed a real-time scheduler to achieve tighter bounds, thus squeezing better performance than traditional real-time operating systems from the same components. However, due to dynamic programming, the method is suitable only for small embedded applications.

Eles et al. presented an approach for system level specification and hardware–software partitioning with VHDL [16]. They formulated HW-SW partitioning as a graph partitioning problem and solved it by implementing two iterative heuristics based on SA and TS. They focused on deriving a perfect cost function that should be optimized. The problem of using a linear weighted sum as the cost function is that the weighing array must be appropriate for the problem and for the desired solution [17]. However, it is hard to find an example of best weighing array due to the complexity of co-synthesis problem that needs exploration of all possible solutions.

Shaha et al. used GA for hardware–software partitioning for some simplified solutions [18]. For example, only one software PE is allowed and no synthesis communication link is carried out. The fitness function considers only violations of constraints and does not try to find better choices among the valid solutions.

Areto et al. provided a formal mathematical analysis of the complexity of partitioning problems [4]. They proved that partitioning problems are generally NP-hard and presented some solvable special cases of partitioning problems. These co-synthesis techniques presented by Areto et al., Prakash and Parker, and Shaha et al. [4,10,18] did not consider real-time constraints.

Various researchers investigated the co-synthesis of multiprocessor embedded systems with real-time constraints. Axelsson compared the three heuristic techniques (TS, SA, and GA) for co-synthesis of real-time systems [19]. Dave and Jha developed a constructive co-synthesis algorithm (COHRA) to solve the co-synthesis problem for multirate hierarchical distributed embedded systems [20]. COHRA supports pipelining of task graphs and employs a combination of preemptive and non-preemptive scheduling algorithms.

Karkowoski and Corporall presented a design space exploration method for homogeneous PEs on a single chip. Their method employs functional pipelining [21]. Oh and Ha designed an iterative algorithm for co-synthesis of system-on-chip devices [22]. Hou and Wolf presented a process clustering method to achieve better co-synthesis [23]. Li and Malik attempted to analyze the extreme (best and worst) case timing bounds of a program on a specific software PE [24].

Real-time constraints for allocation and scheduling problems in multiprocessor embedded systems have been studied by various groups [25,26,27]. Xu presented a scheduling algorithm that solves the problem of finding non-preemptive schedules for systems with identical PEs [25]. However, the identical PEs assumption is not realistic for modern multiple PE embedded systems.

Peng et al. presented a method that finds an optimal solution for allocation of communicating periodic tasks to heterogeneous PE nodes in a distributed real-time system [26]. The methods presented by Xu and Peng et al. use a branch-and-bound search algorithm to find optimal scheduling for target systems. The branch-and-bound method is capable of finding an exact optimal solution and is suitable for small systems with limited numbers of tasks and processes.

Ramamritham presented a static allocation and scheduling algorithm for periodic tasks of distributed systems [27]. The latest start time (LST) and maximum immediate successor first (MISF) heuristics are employed to find feasible allocations and schedules. The priorities of processes are based on their LSTs and the numbers of successor processes. This assignment method does not consider priority levels of the successors. For these methods, only constant numbers of PEs are considered in a multiprocessor system. Therefore, the allocation and scheduling problem has the sole objective of meeting real-time constraints. It makes optimization of these systems much simpler than the process for multiobjective optimization of complex embedded systems.

Wolf presented an iterative architectural co-synthesis algorithm for heterogeneous multiprocessor embedded systems [28]. The algorithm starts by each process to one PE that can execute the process efficiently. To minimize resource cost, and processes are reassigned based on PE utilization, i.e., by removing less utilized PEs from the architecture. In the next stage of the algorithm, process reassignment is performed to minimize PE communication cost. In the final stage of the algorithm, communication channels are allocated among PEs. Depending on whether communication allocation on an existing channel is possible, the existing channel is used or a new one is added.

2.2.4 Co-Synthesis of Real-time Embedded Systems with Multiple PEs

A number of researchers have considered hardware–software co-synthesis with real-time constraints [7,17,29]. Dave et al. presented a new technique (COSYN) for HW-SW co-synthesis of periodic task graphs with real-time constraints [29]. COSYN can produce a feasible distributed embedded architecture for real-time systems. It allows task graphs in which different tasks have different deadlines. They presented a co-synthesis method to optimize resource cost and another co-synthesis method (COSYN-LP) to optimize power. COSYN is efficient for systems in which large numbers of processes are executable on the same type of PEs; clustering the processes speeds process assignment. Clustering of tasks becomes less advantageous for systems with large numbers of PEs whose tasks are executable on different types of PEs.

The multiobjective genetic algorithm (GA) for co-synthesis (MOGAC) uses a multiobjective optimization strategy for hardware–software co-synthesis [7,17]. Dick and Jha presented a co-synthesis method that partitions and schedules embedded system specifications consisting of periodic task graphs [17]. MOGAC caters to real-time heterogeneous distributed architectures and meets hard real-time constraints. It employs a multiobjective GA-based technique to optimize the conflicting features of price and power. MOGAC was designed to accept a database that specifies the performance and power consumption of each task for each available PE type. Each task graph edge is assigned to a communication link after considering power consumption and communication time.

Xie and Wolf presented an allocation and scheduling algorithm for systems with control dependencies among the processes [30]. Their proposed method handles conditional executions in multirate embedded systems. A new mutual exclusion technique for the conditional branches of the task graph is used to exploit resource sharing. The authors employed a scheduling method similar to Sih and Lee's proposed compile-time scheduling heuristic [31].

Sih and Lee used dynamic level scheduling that accounts for interprocessor communication overhead when mapping processes onto a heterogeneous processor architecture. A simplified non-preemptive scheduling technique is used where low priority processes may hamper a higher priority process execution.

Chakravarty and others used a stochastic scheduling algorithm to schedule the processes for their ESCORTS co-synthesis methodology [32]. They used a hierarchical GA-based technique for resource optimization. The GA chromosomes are evaluated via a cost feasibility function. Chromosomes with fitness values below a certain threshold are removed immediately, and the invalid solutions have no chance to evolve into better solutions.

Lee and Ha proposed a co-synthesis method applicable to multiprocessor embedded systems with diverse operating policies [33]. Their method separates partitioning from schedulability analysis and is thus adaptable to various scheduling and operating policies. The method also adopts the schedulability analysis of the timed multiplexing model in the performance evaluation step. Preemptive scheduling is used and an error is reported for any task that cannot meet the real-time constraints. Lee and Ha's method supports only task level deadlines and does not support process level deadlines that may be critical for hard real-time systems [34].

Only a few research projects on multimode embedded systems have been reported. Oh and Ha proposed techniques for multimode, multitask embedded systems with real-time constraints [35]. Their iterative co-synthesis method has three main steps for solving sub-problems separately. A set of PEs is allocated and then the co-synthesis algorithm schedules the acyclic graph of each task to a selected PE to minimize schedule length.

In the final stage of co-synthesis, evaluation is performed to ensure design constraints are satisfied and compute the utilization factor. The drawback is that each task is scheduled independently. To reduce schedule length, all the tasks try to utilize the fastest candidate PE. However, it may cause the fastest PEs to be over-utilized and the slower ones to be under-utilized. Oh and Ha's method assumes that processes in different tasks cannot be executed in parallel and consequently the schedule length becomes unnecessarily longer.

Kim and Kim propose hardware–software partitioning for multimode embedded systems [36]. Unlike Oh and Ha's method [35], they considered process level resource sharing and parallelism rather than task level sharing. Processes are scheduled so that tasks can be executed fully, partially parallel, or completely non-parallel, depending on the status of resource sharing among the processes. The authors also propose a global optimization method for PE allocation for all the tasks and all the modes of a system. Their method schedules processes and each process starts executing as soon as it is ready. Consequently, a lower priority process may be executed while a higher priority process becomes ready, thus hampering the higher priority process execution. This simple scheduling method is not realistic for critical real-time systems.

Process assignment methods for the multimode–multitask co-synthesis methods discussed above, are not efficient for systems with large numbers of candidate PEs [33,35,36]. For each process assignment, every PE is considered. For a large number of PEs, the repeated reassignment and rescheduling may be laborious and sometimes not feasible. Moreover, some of the proposed co-synthesis methods discussed earlier do not provide allocation or scheduling methods for critical communication links. Allocation and scheduling of communication links are essential for system optimization and accurate evaluation of target system architecture. Most co-synthesis methods presented in this section use conventional (and inefficient) task level prioritization for scheduling that may not satisfy real-time process level deadlines.

2.3   CO-SYNTHESIS FRAMEWORK FOR REAL-TIME EMBEDDED SYSTEMS

In this section we describe our co-synthesis framework used for hardware–software partitioning and process scheduling of real-time systems. The following assumptions were made for the target real-time embedded architecture:

- The target system consists of multiple software and hardware PEs that can work in parallel.
- A communication resource supports communication for multiple PEs based on the number of contacts available at the communication resource.
- PEs can perform computations while transferring data to other PEs.

The goal of our co-synthesis method is to produce a suboptimal real-time architecture by allocating PEs, finding process assignments to the PEs, and determining a feasible scheduling order for the architecture. The target architecture should be of low cost and meet the real-time constraints. Our method requires embedded applications in the form of DAGs and a PE library describing the types of processing elements and communication resources. As noted earlier, hardware–software partitioning and scheduling problems are NP-complete and we propose a heuristic optimization method that can generate suboptimal solutions. This method can avoid traps in the local minima and assimilate problem-specific heuristics into the optimization framework.

2.3.1   TARGET EMBEDDED SYSTEM SPECIFICATION AND SOLUTION REPRESENTATION

The proposed co-synthesis methodology accepts a PE library that specifies the execution times of all the processes on each available PE, details of processes that are not executable on a PE, and the cost for each PE. The types of PEs available for a process are represented by an array of integers, and the types of communication resources are represented by a similar array of integers.

For example, the DAG system and partial PE library in Figure 2.3 indicate that a communication resource composed of three contacts and three PEs is available. Moreover, PE0 and PE2 can execute process P1; all three PEs are suitable for executing processes P2 and P3, and process P4 is executable only on PE1.

Process assignment is presented as a one-dimensional array. The offset in this array corresponds to the process. The integer at each offset represents the PE where a process is assigned. For example, for the DAG and PE allocation in Figure 2.3, one possible PE assignment and communication resource assignment solution is presented in Figure 2.4.

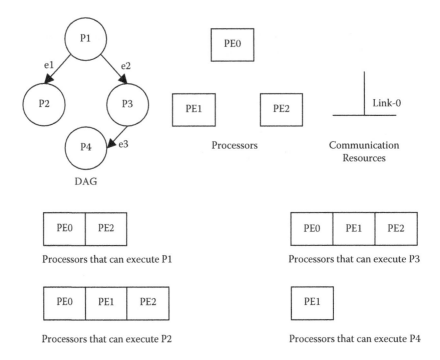

FIGURE 2.3 Processing element and communication resource allocation.

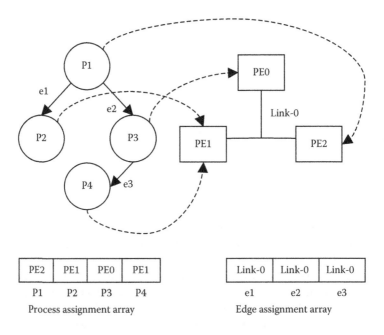

FIGURE 2.4 Processing element and communication resource assignment.

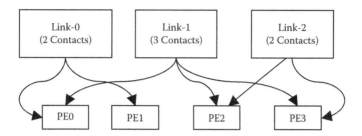

FIGURE 2.5   Communication resource connectivity array.

Every process is assigned to one of the available PEs and Link-0 connects all the three PEs required for data transfer. For every communication resource, an array of PEs specifies the PEs to which the communication resource will be connected. Figure 2.5 shows that Link-0 has two contacts and connects PE0 and PE1. Link-1 has three contacts and connects PE0, PE2, and PE3, and Link-2 has two contacts that connect PE2 and PE3.

From the implementation view of our co-synthesis framework, four main structures model the system. A *node* represents processes, as *edge* represents data communication between processes, *PE* denotes processing elements, and *Link* is for communication resources. Some fields of these structures such as the system specification consisting of DAG, available PEs, and system constraints in the form of cost (hardware area), and real-time process deadlines are specified by the user.

2.3.2   OPTIMIZATION AND PROPOSED CO-SYNTHESIS METHOD

The main steps of our co-synthesis methodology are presented in Figure 2.6. The shaded boxes represent the main GA components. In the first step, the solution pool for the GA is initialized with a random solution. Then GA execution enters the main loop and the other steps of co-synthesis are repeated until the halting condition is reached.

The PE assignment solutions for the generations of GAs are represented by a two-dimensional array in which the first and second dimensions correspond to the solution and the process indices, respectively. The GA chromosome is encoded as an array and the genes as integers as shown in Figure 2.7. For a specific run of this system, process assignment arrays for the first, second, and *nth* solutions of a generation are presented.

2.3.3   PE INITIALIZATION AND COMMUNICATION RESOURCE ALLOCATION

Initial solutions are generated by a simple PE and communication resource allocation method. An available PE capable of executing the first process is chosen randomly. For other processes, a check is made to determine whether one of the already allocated PEs is capable of execution. If none of the allocated PEs is suitable for executing the process, the process is assigned to a new PE randomly. If several already allocated PEs are suitable for executing the process, a randomized selection method will choose a PE to execute the process. The initialization procedure confirms the assignment of all processes to valid PEs at minimum resource cost.

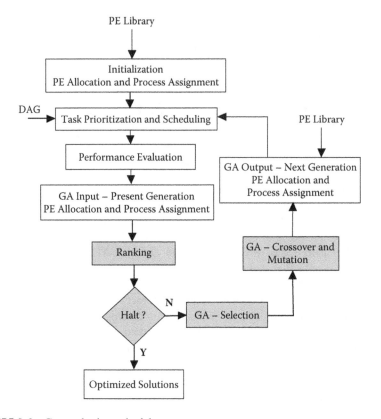

FIGURE 2.6 Co-synthesis methodology.

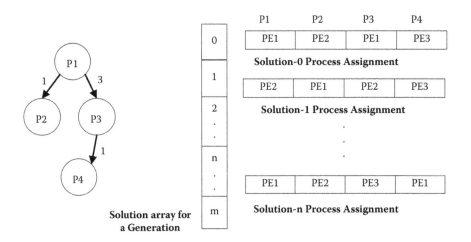

FIGURE 2.7 Solutions for generation.

```
1    for each edge e_i
2        while number of available allocated CR, that have not been checked yet > 0
3            Randomly choose one of the allocated communication resource CR_i
4            if the source process of e_i is one of the contacts of CR_i
5                if the sink process of e_i is one of the contacts of CR_i
6                    Assign e_i to CR_i
7                    break ;
8                else
9                    Use one of the available contacts of CR_i for e_i sink process
10                   Assign e_i to CR_i
11               endif;
12           else
13               if the sink process of e_i is one of the contacts of CR_i
14                   Use one of the available contacts of CR_i for e_i source process
15                   Assign e_i to CR_i
16               endif;
17           endif;
18           if e_i is assigned to CR_i
19               break;
20           else
21               Check another randomly chosen already allocated CR
22       endwhile;
23       if e_i is not assigned to any one of the allocated CR
24           Randomly choose a CR for e_i from the library
25       endif;
26   endfor;
```

FIGURE 2.8 Communication resource (CR) allocation.

Every communication event is assigned to a communication resource. A communication resource can support communication between a limited numbers of PEs, depending on the number of its contacts. The first communication event is assigned to a resource randomly. For other communication events, the already allocated resources are considered first. If one (two) of the contacts of one of the available allocated resources has been used for one PE under consideration (or both), the PE has no need to allocate a new contact PE.

Communication event assignment is performed in such a way that the fewest new communication resources will be added to the architecture. This assignment technique avoids the addition of extra contacts to system resources and minimizes total resource cost. If none of the already used communication resources appears economical, a new resource is randomly selected to perform the communication event. Figure 2.8 depicts a communication resource (CR) allocation algorithm.

2.3.4 DEADLINE ASSIGNMENT

The hyperperiod of the system is the least common multiple (LCM) of the periods of tasks. Each task has *hyperperiod/period* copies for the duration of the hyperperiod. If all the copies of all tasks of a target system are schedulable within the hyperperiod, the system is schedulable for the solution [33]. The proposed method considers deadlines longer than the period. This means multiple copies of the same task can be ready to execute simultaneously. Deadlines for the first copy of each task are

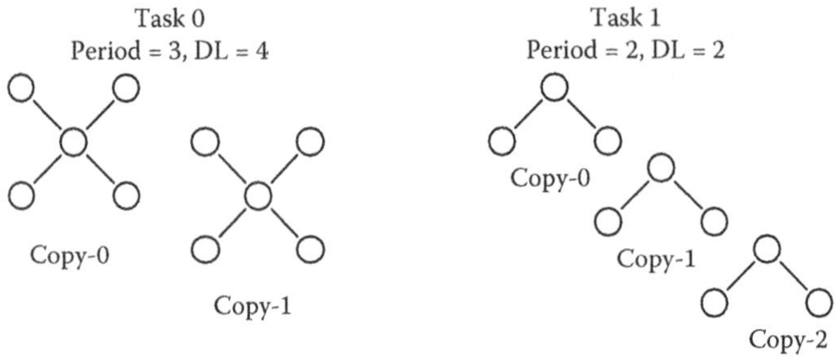

FIGURE 2.9 Deadlines and periods of two-task system.

specified by the user. The proposed method also supports process level deadlines. Deadlines for the copies of a process are calculated as shown in Equation (2.2).

*Deadline of nth copy of process = Process period * (n – 1) + Deadline of 1st copy* (2.2)

For example, consider a two-task system as shown in Figure 2.9 with a hyperperiod of 6. Task 0 has a period of three and two copies of task 0 should be executed in the hyperperiod. Task 1 has a period of two and three copies should be executed in the hyperperiod. The first copy of the sink node of task 0 has a deadline of four. The second copy of task 0 is ready to be executed at time unit three while the first copy might still be in the process of execution.

The second copy of task 0 has a deadline of seven. The first copy of the sink node of task 1 has a deadline of two, and its second copy is ready for execution at time unit two with a deadline of four. The third copy of task 1 is ready at time unit four and has a deadline of six.

2.3.5 PROCESSES AND COMMUNICATION EVENT SCHEDULING

During co-synthesis, scheduling is carried out to evaluate each solution. Our static scheduling method is non-preemptive and allows preemption only for special cases in which preemption time is smaller than schedule length. This will result in lower context switch overheads compared to traditional preemptive scheduling.

The processes of an input task are scheduled on their assigned PEs in the order of priority. This prioritization method makes it possible to meet hard real-time constraints. The data transfer time is added to the parent process end time to calculate the ready time of a process when parent and dependent process are allocated to different PEs.

2.3.5.1 Process Prioritization

During each GA generation and for each GA solution, all copies of all the processes are prioritized relatively based on deadline, execution time, and children priorities after the processes are assigned to PEs. If a process has no deadline, the shortest deadline of its child processes is considered the deadline.

Unlike many other conventional prioritization methods [33,35,36], our technique assigns priorities at process rather than task level, even when deadlines are not specified at the process level. This guarantees that the ancestors of a sink process complete their execution in time and the sink process will meet its deadline. The priority assignment methodology used in our co-synthesis procedure is an improved version of the priority assignment employed in COSYN [29].

We use an application task graph, PEs, and communication resources (Figure 2.10) to explain the prioritization method. PE and the communication resource libraries appear in Tables 2.1 and 2.2, respectively. We also assume that a solution is generated by the co-synthesis framework for which process and communication event (edge) assignments are shown in Figure 2.11. Execution time on the assigned PE is given for each process shown in parentheses below the process assignment array.

Similarly with the communication event assignment array, data transfer time is presented for each edge in parentheses. Data transfer time for a communication event is calculated by using Equation (2.3). The data transfer time between a source and destination process is calculated by employing the amount of data to be transferred between the processes and the transfer rate of the communication resource to which the event is assigned. Priority level assignment for the processes can be determined by Equations (2.4) and (2.5).

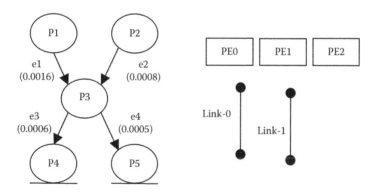

FIGURE 2.10 Task graph, processing elements, and communication resources.

TABLE 2.1

Processing Element Library

	Execution Times		
Process	PE0 (cost: 40)	PE1 (cost: 20)	PE2 (cost: 35)
1	0.7	1.5	1
2	0.5	1.7	1.3
3	–	1	–
4	1	–	0.7
5	1.3	–	0.5

TABLE 2.2

Communication Resource Library

Link	Cost per Contact	Number of Contacts	Rate
0	6	2	1
1	10	2	0.8

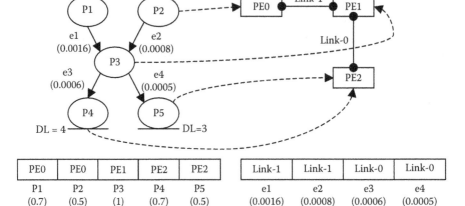

PE0	PE0	PE1	PE2	PE2		Link-1	Link-1	Link-0	Link-0
P1	P2	P3	P4	P5		e1	e2	e3	e4
(0.7)	(0.5)	(1)	(0.7)	(0.5)		(0.0016)	(0.0008)	(0.0006)	(0.0005)

FIGURE 2.11   Process and edge assignment.

$$Data\ transfer\_time\ of\ edge = Communication\_resource\_rate * amount\_of\_data \tag{2.3}$$

$$Priority\ level\ of\ sink\ process = Proc\_time - Deadline \tag{2.4}$$

$$Priority\ Level\ of\ non\_sink\ process = Proc\_time + max\ (Child\_Priority\_level - Deadline) \tag{2.5}$$

The priority of a process is assigned by employing the above priority level equations. For the process and communication event assignment in Figure 2.11, priority levels and priorities of all the processes are presented in Table 2.3.

Proc_time is the execution time of a process for the assigned PE and also includes data communication time (maximum of data transfer times for all incoming edges). For example, to calculate the *Proc_time* for process P3, the maximum data transfer time (0.0016) for edge *e1* is added to the execution of P3. The *Proc_time* of a process differs from solution to solution, depending on the PE assigned and the communication resource on which incoming edges are assigned.

Child process priority for a process is treated as the maximum priority of all its dependent processes. For example, child priority for P3 in Table 2.3 is the maximum

TABLE 2.3

Priority Assignment

Process	Proc_Time	Child Priority	Deadline	Priority Level	Priority
P1	0.7	−1.4979	3	−0.7979	5
P2	0.5	−1.4979	3	−0.9979	4
P3	1.0016	−2.4995	3	−1.4979	3
P4	0.7006	−	4	−3.2994	1
P5	0.5005	−	3	−2.4995	2

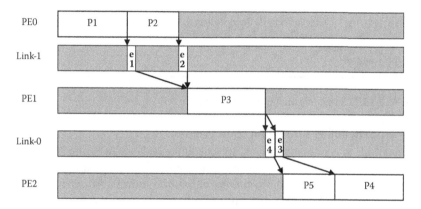

FIGURE 2.12   Example of PE and CR schedules.

of P4 priority (−3.2994) and P5 priority (−2.4995). Deadline (DL) in the fourth column of the table denotes the deadlines of processes. Priority levels of the processes are calculated by employing Equations (2.4) and (2.5). In this case, P1 has the highest priority and P4 has the lowest.

The prioritization method helps meet real-time constraints and achieves better results that prioritization methods used in rate monotonic and deadline monotonic scheduling [37,38]. In deadline monotonic scheduling, priority assignment is based on process deadline alone. If a process has a deadline two times longer than that for a second process and has an execution time three times longer than that of the second process, the first process has a high probability of missing its deadline. With deadline monotonic scheduling, the first process will have lower priority than the second. With our method, the second process will be prioritized higher.

Priorities of processes with children depend on child priority as well. The priority of a process differs from one solution to another because priorities depend on process execution times. Communication events are also prioritized and a communication for a higher priority destination process receives a higher priority. The schedule lists for the PEs and communication resources based on the priorities in Table 2.3 are depicted in Figure 2.12.

```
1     for each process in the order of priority
2          for each incoming edge
3               EdgeReadyTime = SourceProcessEndTime;
4               PresentTime = EdgeReadyTime;
5               for each HP edge in the order of scheduling sequence of CRᵢ
6                    if PresentTime <HP EdgeStartTime
7                         if (PresentTime + DataTransferTime) <=HP EdgeStartTime{
8                              The Edge is scheduled in this empty slot;
9                              break;
10                        } else
11                             PresentTime = HP EdgeEndTime;
12                        if (HP EdgeStartTime <PresentTime <HP EdgeEndTime)
13                             PresentTime = HP EdgeEndTime;
15               endfor;
16               EdgeStartTime =PresentTime;
17               ParentDataSentTime =EdgeStartTime +DataTransferTime;
18          endfor;
19     endfor;
```

FIGURE 2.13   Communication event scheduling.

2.3.5.2   Scheduling Communication Events

A pseudocode of communication events scheduling is shown in Figure 2.13. HP indicates higher priority and CR_i is the assigned communication resource for an edge i. Communication events for the incoming edges of a process are scheduled on the assigned resources. After the source process of an edge completes execution, a communication event for the edge can take place on the assigned resource.

To schedule a communication event on a communication resource, data transfer start and end times for all scheduled events are checked. If the communication event can start and finish data transfer before one of the scheduled communications is ready for transfer, it is scheduled for the empty slot (lines 6, 7 and 8). If no suitable empty slot is available among the higher priority edges, the edge must wait until all the higher priority edges complete data transfer on the communication resource. The destination process of an edge receives the required data from the source process at *ParentDataSentTime*.

2.3.5.3   Scheduling of Processes

Figure 2.14 presents the process scheduling algorithm for the PEs. HP means higher priority and PE_i indicates the PE on which the process is scheduled. A process is ready to execute when required data from its parent processes are received. Process execution can take place on the assigned PE when it has an empty time slot to execute the process.

To schedule a process on a PE, execution start and end times for all the scheduled processes are checked. If a process can complete execution before one of the scheduled processes is ready to execute, it is scheduled for the empty slot of the PE (lines 5 and 6). In the case of preemptive scheduling, part of a process execution is allowed in an empty slot only if the preemption time is smaller than the empty slot (lines 9 and 10). The finish time of a process is calculated by adding process start time to the execution time for the process.

```
1    for each process in the order of priority
2          ProcessReadyTime = max of all ParentDataSentTime ;
3          for each scheduled HP process in the order of scheduling sequence of the PEᵢ ;
4                EmptyTimeSlot = HP ProcessStartTime – ProcessReadyTime ;
5                if (EmptyTimeSlot> 0) and (ProcessExecutionTime <= EmptyTimeSlot) {
6                      The Process is scheduled in this empty slot ;
7                      break ; /* No need to check other HP processes */
8                } else
9                      if (Preemption is allowed) and (ProcessPreemptionTime < EmptyTimeSlot) {
10                           A portion of the Process is scheduled in the empty slot ;
11                           ProcessStartTime = ProcessReadyTime ;
12                           ProcessExecituiomTime –= EmptyTimeSlot ;
13                           ProcessExecituiomTime += ProcessPreemptionTime ;
14                           ProcessReadyTime = HP ProcessFinishTime ;
15                      } else /* no empty slot before the HP process */
16                           ProcessReadyTime = HP ProcessFinishTime ;
17          endfor ;
18          if the process was not preempted
19                ProcessStartTime = ProcessReadyTime ;
20          ProcessFinishTime = ProcessReadyTime + ProcessExecutionTime ;
21   endfor ;
```

FIGURE 2.14   Process scheduling.

2.3.5.4   Better PE Utilization

The proposed scheduling technique allowed better PE utilization than conventional non-preemptive scheduling. We assume that if all the copies of tasks are schedulable within the hyperperiod, the tasks are schedulable because they will satisfy schedulability conditions [33]. For conventional scheduling, higher priority processes are always scheduled on a PE before low priority processes. As a result, many unused time slots on PEs may be used to execute lower priority processes that are ready for execution.

Our scheduling method allows a low priority process execution before the higher priority processes if the higher priority processes can still be completed before their deadlines. In this way, higher priority process completion (before deadline) is not hampered by execution of a lower priority process.

We compared the utilization factors for both methods by using Equations (2.6) to (2.8). The processes under consideration are assigned to PE_a. We assume n is the priority of the highest priority process and $(n - y)$ is the priority of a process that is the last to be ready for execution in the hyperperiod (due to data dependency). Moreover, another with a priority of $(n - z)$ exists; y and z are integers such that $(z > y)$. The $(n - z)$ priority process has lower priority than the $(n - y)$ process.

Let us also assume that process with priority $(n - z)$ is ready to execute before some higher priority processes. Execution time for processes with priority n is denoted by c_n and execution time for the processes with priority less than n is denoted by $c_{n-1}, c_{n-2} \cdots$ and so on. An empty time slot between the finish time of a process with priority n and ready time of a process with priority $n - 1$ is large enough to execute a process with priority $(n - z)$. Equation (2.6) presents the PE utilization for conventional scheduling. Equation (2.7) determines processor utilization for our proposed scheduling technique.

$$PE_a\_utilization(conv.) = (c_n + c_{n-1} + c_{n-2} + \dots + c_{n-y})/hyper\_period \qquad (2.6)$$

$$PE_a\_utilization(prop.) = (c_n + c_{n-z} + c_{n-1} + c_{n-2} + \dots + c_{n-y})/hyper\_period$$
$$(2.7)$$

$$PE_a\_utilization(prop.) - PE_a\_utilization(conv.) = c_{n-z}/hyper\_period \quad (2.8)$$

We can observe from Equation (2.8) that the proposed method achieves better utilization than conventional methods by a factor of $c_{n-z}/hyper\_period$. We use the E3S benchmark for telecommunication to compare the PE utilization for our scheduling with the conventional method. Details of E3S benchmarks are presented in Section 2.4.4.

Consider an embedded system solution with three PEs (PE1, PE5, and PE13) for a resource cost of 297 (including communication resources). The hyperperiod for this system is 0.001 time units. Figure 2.15 presents the schedule for the processes of the system assigned to PE13 by using conventional non-pre-emptive scheduling. Note from Figure 2.15 the large empty slot on this PE between the processes with priorities 34 and 35. All the lower priority processes are waiting for the priority 34 process to complete its execution.

As a result of the waiting, the schedule length for the processes is extended to 0.00152 time units. PE13 utilization for the hyperperiod is 0.398. The total schedule length exceeds the hyperperiod, indicating that the processes assigned to PE13 are not schedulable and this architecture is not a valid solution. If we explore the design space with all the other options of process assignment, we can conclude that no valid solution can be produced for a 297 cost by employing conventional scheduling.

Figure 2.16 presents a schedule for the system processes assigned to PE13 based on our proposed method. We can see from the figure that by allowing lower priority processes to execute before higher priority processes, the empty slots for PE13 are filled and PE13 utilization for the hyperperiod is increased to 0.926. Using this technique, it is now possible to generate a valid schedule (shorter than hyperperiod 0.001) for lower resource cost. Additional details of the scheduling experiments are available elsewhere [39].

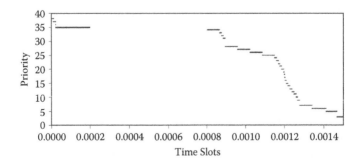

FIGURE 2.15   Schedule for conventional non-pre-emptive method.

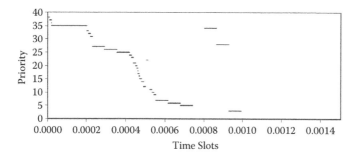

FIGURE 2.16 Proposed scheduling method.

2.3.6 EVALUATION OF ARCHITECTURAL CO-SYNTHESIS

The evaluation of solution architecture is performed by calculating the system cost, total system resources (hardware area) used, application execution time, and number of deadline violations. The completion time of each process is recorded during scheduling. We can verify whether a process missed its deadline by inspecting its completion time. Every task of a system has one or more processes with specified deadlines.

Hard real-time constraint violation for a system is the sum of deadline violations of such processes in all the task copies of the system. Resource cost for a solution is determined by summing the cost of all the PEs and communication resources used by the architectural solution. A solution is invalid if it meets any of the following conditions:

- Violation of resource constraints
- Failure to meet specified deadline
- Lack of required communication links for target architecture
- Non-schedulable processes or tasks.

Invalid architectural solutions are not terminated immediately. It may be possible to repair invalid architectures but it is difficult to formulate a repair operation guaranteed to produce only valid repair solutions [7]. Therefore, invalid solutions are treated the same way as valid solutions during evolution. Invalid solutions are given a chance to reproduce for the next generations and evolve into high-quality valid solutions by mutation and crossover.

Valid solutions may violate soft real-time constraints. It is desirable to reduce the cost or schedule length until it is lower than its soft real-time constraint. Our co-synthesis methodology performs solution evaluation for every solution of each generation.

2.3.7 EVOLUTION OF NEW SOLUTIONS

The number of solutions for every generation is constant during a run of a GA. This number is chosen at the start of a GA run. For our co-synthesis method, this number depends on system size and complexity.

2.3.7.1 Solution Ranking

The invalid solutions are ranked based on the extent by which they violate the hard real-time constraints. For valid solutions, ranking is based on the resource cost. A lower resource cost of a solution leads to higher rank. If two valid solutions have the same resource cost, ranking is based on soft deadline violations and schedule length. The solutions with better (smaller) deadline violation levels and shorter schedule lengths are ranked higher.

2.3.7.2 Halt

After ranking and if halting conditions have not yet been reached, the solutions are subjected to a selection process for the next generation evolution. These steps are repeated until one of the halting conditions is reached. A halting condition occurs when multiple generations pass without a change in the solution pool. At the end of a GA run, the feasible solutions of the final generation that meet the imposed constraints are presented to the user. These solutions present the target system design space with trade-off points.

2.3.7.3 Solution Selection and Reproduction

During selection, some architectural solutions in the present generation are selected for reproduction. The higher the rank of a solution, the more likely its selection for reproduction. A good solution produced for a generation may be lost in the next generation by random selection, mutation, or crossover. That is why a best solution for a generation that meets all the constraints is passed on to the next generation without changes. In this way, the best ranked solution is preserved until a better one evolves in another generation and is then ranked as the best solution. The solution selection algorithm used in the GA-based co-synthesis method appears in Figure 2.17.

2.3.7.4 Mutation and Crossover

Mutation and crossover operations are performed on the selected PE assignment arrays. For a randomly selected process (during mutation), the PE assignment is changed by another PE that is capable of executing the process. During crossover operation, two offsets on the arrays are chosen randomly for a randomly selected

```
 1  for total number of solutions process in the order of priority
 2      Generate a random number between 1 and Total_Rank ;
        /* Total_Rank = sum of ranks of all the processes * /
 3      for each solution Sᵢ ;
 4          NextRank = PreviousRank + The solution rank;
 5          if (PreviousRank <= random number < NextRank) {
 6              The solution Sᵢ is selected ;
 7                  break ; /* No need to check other HP processes */
 8          } else
 9              PreviousRank  = NextRank;
10  endfor ;
11  endfor ;
```

FIGURE 2.17 Solution selection.

pair of PE assignment arrays. Integer-presenting PE types for the offsets between the randomly selected offsets are exchanged between the selected arrays.

For example, consider the assignment arrays shown in Figure 2.7. We assume assignment arrays for solutions 0 and 1 are randomly selected for crossover. The randomly chosen offsets on these arrays are numbered 2 and 3. Therefore PE assignments are exchanged for offsets 2 and 3 between PE assignment arrays 0 and 1. For the new solution 1, process 2 will be assigned to PE1 and process 3 will be assigned to PE2.

To preserve the locality of the solutions toward the end of a GA run, the probabilities of crossover and mutation are lowered for the later generations. After these crossover and mutation operations, newer generations of hardware–software partition solutions evolve and are prioritized, scheduled, evaluated, and ranked as described earlier. The GA-related algorithms for crossover and mutation are presented in Figures 2.18 and 2.19.

```
1   for every generation gᵢ
2        for i = 0 to (TotalNumber_of_Solutions)/2
3             Randomly choose one pair of solutions (PE assignment arrays) ;
4             if gᵢ < (TotalNumber_of_Generations)/2
5                  probability of crossover is 1/2;
6             else probability of crossover is 1/5
7             if the pair is selected for crossover {
8                  randomly choose two offsets(processes) of the solutions ;
9                  swap the integers(PEs) at each off set between these offsets;

10                }
11  endfor ;
12  endfor ;
```

FIGURE 2.18   Crossover algorithm.

```
1   for every generation gᵢ
2        for every solution (PE assignment array)
3             for every offset (process) of the solution
4             if gᵢ < (TotalNumber_of_Generations)/2
5                  probability of crossover is 1/2
6             else probability of crossover is 1/5
7             if the offset (process) is selected for mutation {
8                  randomly choose a PE that can execute the process ;
9                  change the integer (old PE) at the offset with the new PE;
10                }
11             endfor ;
12        endfor ;
13  endfor ;
```

FIGURE 2.19   Mutation algorithm.

2.4 EXPERIMENTAL RESULTS

The co-synthesis method for real-time embedded systems was implemented using C/C++ language and results obtained by execution on an UltraSPARC workstation with 4 Gb of memory in a Linux environment. The method was tested with different randomly generated graphs and some well-known benchmarks. In this section, we present these results and compare them with the results for earlier techniques proposed in the literature.

2.4.1 MPEG-2 ENCODER

For experimental purposes, an MPEG encoder application was used as an input to our co-synthesis technique. The directed acyclic graph (DAG) with 21 coarse-grained processes and PE library used for the system were adapted from one of our hardware–software co-synthesis methods for multiprocessor systems [40]. A simplified version of the MPEG-2 encoder DAG is given in Figure 2.20.

A comparison between the first generation of random solutions and the final generation of optimized solutions for our GA based co-synthesis for the MPEG encoder system, is plotted in Figure 2.21. The schedule length (performance) of the system is in clock cycles. These results indicate that our co-synthesis algorithm can produce an optimized solution with 85% better performance (shorter task completion time) as compared to the first generation random solution with the same area cost (8963 logic block units). For the same schedule length (30 million time units), our co-synthesis approach for real-time systems produced an optimized architecture with 18% lower area cost as shown in Table 2.4. The co-synthesis method execution time for a complex 21-node task graph was 4.08 seconds.

2.4.2 HOU AND WOLF'S TASK GRAPHS

We also tested our co-synthesis method using Hou and Wolf's example task graphs [23]. Figure 2.22 shows the DAGs for the Hou and Wolf's graphs. Table 2.5 presents the PE library for these tasks and shows PE execution time for each process. The values in parentheses associated with the PEs represent cost. The cost for a point-to-point communication link is 20.

Table 2.6 compares the results produced by our co-synthesis algorithm for Hou and Wolf's task graphs with results cited in the literature. The first column lists Hou and Wolf's examples, for example, H&W 1,2 refers to a multitask system consisting of tasks 1 and 2 in Figure 2.22. The second column shows the cost of the solution produced by the methodology when Hou and Wolf's original task deadlines are used.

The third column shows the cost of system solutions produced by our method when deadlines were assumed to be the same as the periods of the tasks. The fourth column lists results produced by Yen and Wolf's co-synthesis algorithm [41], the fifth column shows COSYN results [17], the sixth column shows EMOGAC results [7], and final column lists results from the co-synthesis algorithm of Lee and Ha [33].

Lee and Ha assumed that the task deadlines were the same as the task periods [33]. When the deadline was relaxed to be the same as the period of a task, our

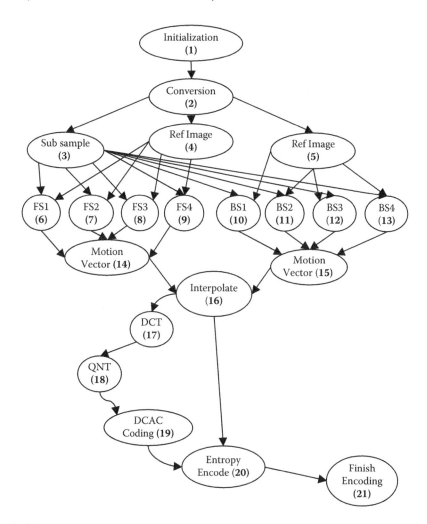

FIGURE 2.20 MPEG-2 encoder DAG.

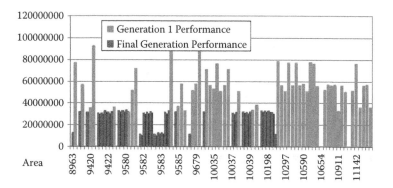

FIGURE 2.21 Initial and final generation solutions for MPEG2 encoder.

TABLE 2.4

MPEG-2 First and Final Generations

Generation	Resource Cost (Schedule Length = 30 Million)
1	10,591 units
Final	8,963 units

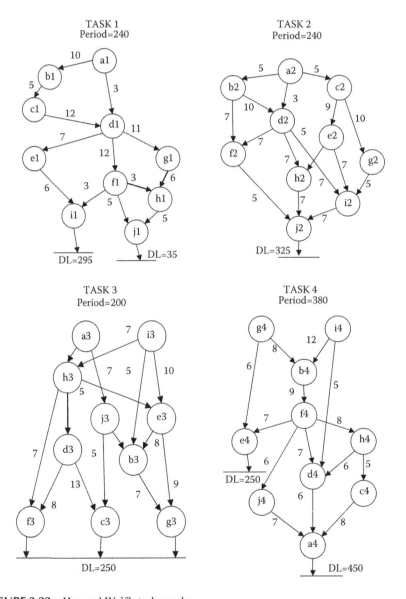

FIGURE 2.22 Hou and Wolf's task graphs.

TABLE 2.5

Processing Element Library: Hou and Wolf's DAG

Process	PEX (100)	PEY (50)	PEZ (20)
a	5	12	18
b	10	18	40
c	5	12	18
d	35	85	95
e	15	22	80
f	30	75	180
g	15	25	85
h	15	35	47
i	7	10	30
j	10	28	35

TABLE 2.6

Results of Hou and Wolf's Example Task Graph

DAG	Proposed Deadline (Hou)	Proposed Deadline (Period)	Yen & Wolf	COSYN	EMOGAC (Hou)	Lee et al. (Period)
H&W 1, 2	100	100	170	170	140	150
H&W 1, 3	170	140	170	170	170	170
H&W 3, 4	140	100	170	–	140	170

---

methodology produced better solutions at lower cost than those produced by past co-synthesis methods for real-time systems. When the deadline was tightened, our method generated solutions at costs equal to or lower than other methods presented in the literature.

2.4.3   E3S BENCHMARKS

In this section, we present the co-synthesis results for E3S benchmarks of four industrial systems: telecommunication, networking, office automation, and automotive [42]. Figure 2.23 shows some of the main tasks of E3S benchmarks for a telecommunication system. The main functional modules (processes) are auto-correlation (ac), convolutional encoder (ce), fixed-point bit allocation (fpba), fixed point complex fast Fourier transformation (fft), and global system for mobile communications (gsm).

The system consists of nine tasks that share these functional modules. Only six of the main tasks (0 through 5) are shown in Figure 2.23. The three tasks not shown are two process tasks. For this system, Table 2.7 and Figure 2.24 demonstrate solution quality improvement by the evolution of GAs for our co-synthesis algorithm run. For

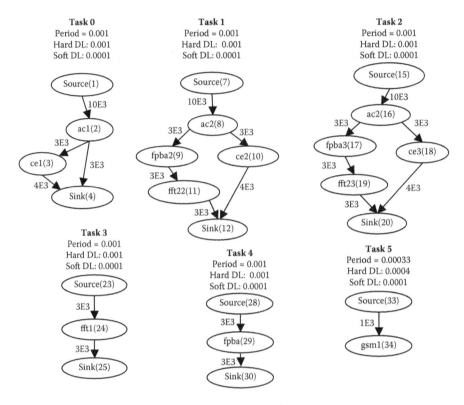

FIGURE 2.23 Telecommunication E3S benchmark.

this run of the GA in our co-synthesis method, 100 generations (iterations) were used and every generation had a pool of 1,000 solutions.

Table 2.7 indicates that for the first generation, the 722nd solution ranked tenth. This solution required a resource cost of 297 and violated the soft deadlines by 0.0692. The amount of soft deadline violation was calculated as the sum of the times by which every copy of every task missed its soft deadline. For the final (999th) generation, solution 976 ranked tenth. It required the same area cost and exhibited a smaller deadline violation of 0.0272 time units.

The best ranked solution for generation 1 showed a resource cost of 284 and violated the soft deadlines by 0.0052 time units. The GA evolution technique greatly improved the quality of the solutions. Furthermore, for the same resource cost (284) for generation 1, the best solution of the final generation missed the soft deadline by only 0.0029 time units. The solution quality improvement by GA evolution for this co-synthesis run is plotted in Figure 2.24.

We also tested our method for the co-synthesis of real-time systems for automotive, networking, and office automation systems. Table 2.8 compares the results produced by our algorithm for E3S benchmarks [42] with EMOGAC [7]. The soft deadline violation proportion is defined as the sum of the times by which every copy of each task misses its soft deadline divided by the hyperperiod. The first column

TABLE 2.7

Solution Quality Improvement by GA Evolution

	1st Generation			999th Generation		
Rank	Solution	Area	Soft Deadline Violation	Solution	Area	Soft Deadline Violation
10	722	297	0.0692	976	297	0.0272
100	369	297	0.0365	356	297	0.0170
200	406	297	0.0297	381	297	0.0115
300	530	284	0.0326	803	284	0.0163
400	813	297	0.0261	814	297	0.0177
500	903	297	0.0252	972	297	0.0093
600	892	297	0.0194	832	297	0.0119
700	102	297	0.0131	190	297	0.0108
800	356	297	0.0158	795	297	0.0072
900	643	297	0.0175	181	297	0.0064
970	211	297	0.0058	536	297	0.0057
990	63	284	0.0055	510	284	0.0049
1000	290	284	0.0052	0	284	0.0029

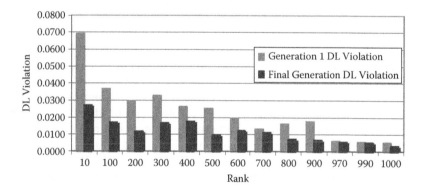

FIGURE 2.24   First and final generations' soft deadline violations.

lists E3S benchmark examples for automotive, telecom, networking, and office automation applications. The second and third columns list the prices and deadline violation proportions presented by Dick [7]. The fourth and fifth columns show the cost and deadline violation proportions of the solutions produced by our framework. Our proposed co-synthesis method achieved better deadline violation proportions at lower resource cost than solutions produced by EMOGAC.

2.4.4   MULTIMODE APPLICATIONS

To test the multimode feature of our co-synthesis method, we employed Hou and Wolf's task graphs from Figure 2.22 [23]. We considered a multimode system.

TABLE 2.8

Results for E3S Benchmarks

Benchmark	EMOGAC		Proposed	
	Cost	Soft Deadline Violation	Cost	Soft Deadline Violation
Automotive	169	2.08	139	1.17
	530	1.05	301	0.53
Telecom	291	4.58	228	3.70
	378	3.18	284	3.07
Networking	57	1.31	52	0.726
	70	1.23	76	0.64
Office Automation	66	0.02	65	0.02

The first mode was an application with Hou and Wolf's tasks 1 and 2. The second mode was an application with tasks 1 and 3. The third utilized tasks 3 and 4. The goal was to determine a suitable globally optimized architecture for real-time use for all three applications.

While PE allocation and process assignment to PEs were performed for a particular mode, all the other system modes were considered as well. We compared the results of our method with the results for Oh and Ha's algorithm [35] and those for Kim and Kim's algorithm [36]. Our method achieved equal or better quality architecture solutions than those methods.

2.5   CONCLUSIONS

In this paper, we presented a co-synthesis methodology for multiprocessor–multitask systems with real-time constraints. The proposed method determines feasible solutions with optimized hardware–software partitioning and real-time schedules for processes and data communication events. The method is capable of producing acceptable solutions for critical systems with hard real-time deadlines by employing process level priorities and by meeting process level deadlines. Previous co-synthesis methods presented in the literature do not provide allocation techniques and scheduling methods for the communication events [29,33] that are crucial for system optimization and accurate solution evaluation.

The presented methodology employs efficient scheduling mechanisms and allocation techniques for PEs and communication resources. Moreover, it is capable of achieving better PE utilization. Our method is accurate and efficient for finding a feasible optimized architecture. It produces low cost architectures for fairly large and complex embedded systems and satisfies schedulability conditions.

Our proposed co-synthesis method can be improved in a number of ways. One interesting direction could be the co-synthesis of dynamically reconfigurable embedded systems. Dynamic reconfiguration may reduce the amount of hardware required in a multimode embedded system. Recently the flexibility and reconfiguration speed of FPGAs have been improved. Therefore, the option of reconfiguring FPGAs while an embedded system operates has become more practical. Finally more non-functional requirements could be considered in addition to system area (cost) and real-time constraints. Reliability is a critical metric for fault-tolerant systems that could be considered a multiobjective optimization goal for co-synthesis methodology. Low power consumption can also be considered a non-functional requirement.

ACKNOWLEDGMENTS

The author acknowledges his graduate students for generating the experimental results presented in this chapter and the financial and equipment support provided by NSERC and CMC Corporation.

REFERENCES

1. S. Chakraborty, M. Di Natale, H. Falk et al. 2011. Timing and schedulability analysis for distributed automotive control applications. In *Proceedings of EMSOFT*, pp. 349–350.
2. M. Kim, S. Banerjee, N. Dutt et al. 2008. Energy-aware co-synthesis of real-time multimedia applications on MPSoC using heterogeneous scheduling policies. *ACM Transactions on Embedded Computing Systems*, 7, 9:1–9:19.
3. O. Cheng, W. Abdulla, and Z. Salcic 2011. Hardware–software co-design of automatic speech recognition system for embedded real-time applications. *IEEE Transactions on Industrial Electronics*, 58, 850–859.
4. P. Areto, S. Juhasz, Z. Mann et al. 2003. Hardware–software partitioning in embedded system design. In *Proceedings of IEEE International Symposium on Intelligent Signal Processing*, pp. 197–202.
5. J. D. Ullman. 1975. NP-complete scheduling problems. *Journal of Computer Systems Science*, 10, 384–393.
6. J. Nestor and D. Thomas. 1986. Behavioral synthesis with interfaces. In *Proceedings of Design Automation Conference*, pp. 461–466.
7. R.P. Dick. 2002. Multiobjective Synthesis of Low-Power Real-Time Distributed Embedded Systems. PhD thesis, Princeton University.
8. A. Tino and G.N. Khan. 2011. Designing power and performance optimal application-specific network-on-chip architectures. *Microprocessors and Microsystems*, 35, 523–534.
9. D. E. Goldberg. 1989. *Genetic Algorithms in Search, Optimization, and Machine Learning*. Reading, MA: Addison-Wesley.
10. S. Prakash and A. Parker. 1992. SOS: synthesis of application-specific heterogeneous multiprocessor systems. *Journal of Parallel and Distributed Computing*, 16, 338–351.
11. R.K. Gupta and G.D. Micheli. 1993. Hardware–software co-synthesis for digital systems. *IEEE Design and Testing of Computers*, 10, 29–41.
12. R. Ernst, J. Henkel, and T. Benner. 1993. Hardware–software co-synthesis for microcontrollers. *IEEE Design and Testing of Computers*, 10, 64–75.
13. H. Henkel and R. Ernst. 1997. A hardware–software partition using a dynamically determined granularity. In *Proceedings of 34th Conference on Design Automation*, pp. 691–696.
14. H. Liu and D. Wong. 1998. Integrated partitioning and scheduling for hardware–software co-design. In *Proceedings of International Conference on Computer Design: VLSI in Computers and Processors*, pp. 609–614.
15. V. Mooney, III and G.D. Micheli. 1997. Real time analysis and priority scheduler generation for hardware–software systems with a synthesized runtime system. *Computer-Aided Design. Digest of Technical Papers*, November.

16. P. Eles, Z. Peng, K. Kuchcinski et al. 1996. Hardware–software partitioning of VHDL system specifications. In *Proceedings of European Design Automation Conference*, pp. 434–439.
17. R. P. Dick and N. K. Jha. 1998. MOGAC: a multiobjective genetic algorithm for the co-synthesis of hardware-software embedded systems. *IEEE Transactions on Computer-Aided Design*, 17, 920–935.
18. D. Shaha, R.S. Mitra, and A. Basu. 1997. Hardware–software partitioning using genetic algorithm. In *Proceedings of Tenth International Conference on VLSI Design*, pp. 155–160.
19. J. Axelsson. 1997. Architecture synthesis and partitioning of real-time systems: a comparison of three heuristic search strategies. In *Proceedings of Fifth International Workshop on Hardware–Software Codesign*, pp. 161–165.
20. B. Dave and N. Jha. 1998. COHRA: hardware–software co-synthesis of hierarchical distributed embedded system architectures. In *Proceedings of 11th International Conference on VLSI Design*, pp. 347–352.
21. I. Karkowski and H. Corporall. 1998. Design space exploration algorithm for heterogeneous multiprocessor embedded system design. In *Proceedings of Design Automation Conference*, pp. 82–87.
22. H. Oh and S. Ha. 1999. A hardware–software co-synthesis technique based on heterogeneous multiprocessor scheduling. In *Proceedings of Seventh International Workshop on Hardware–Software Codesign*, pp. 183–187.
23. J. Hou and W. Wolf. 1996. Process partitioning for distributed embedded systems. In *Proceedings of Fourth International Workshop on Hardware–Software Co-Design*, pp. 70–76.
24. Y. Li and S. Malik. 1995. Performance analysis of embedded software using implicit path enumeration. In *Proceedings of 32nd ACM/IEEE Conference on Design Automation*, pp. 456–461.
25. J. Xu. 1993. Multiprocessor scheduling of processes with release times, deadlines, precedences, and exclusion relations., *IEEE Transactions on Software Engineering*, 19, 139–154.
26. D.T. Peng, K. Shin, and T. Abdelzaher. 1997. Assignment and scheduling communicating periodic tasks in distributed real-time systems. *IEEE Transactions on Software Engineering*, 23, 745–758.
27. K. Ramamritham. 1995. Allocation and scheduling of precedence-related periodic tasks. *IEEE Transactions on Parallel and Distributed Systems*, 6, 412–420.
28. W. Wolf. 1997. An architectural co-synthesis algorithm for distributed, embedded computing systems. *IEEE Transactions on Very Large Scale Integration Systems*, 5, 218–229.
29. B. Dave, G. Lakshminarayana, and N. K. Jha. 1997. COSYN: hardware–software co-synthesis of embedded systems. In *Proceedings of 34th Annual Conference on Design Automation*, pp. 703–708.
30. Y. Xie and W. Wolf. 2001. Allocation and scheduling of conditional task graph in hardware–software co-synthesis. In *Proceedings of Design, Automation and Testing in Europe*, pp. 620–625.
31. G. C. Sih and E. A. Lee. 1993. A compile-time scheduling heuristic for interconnection-constrained heterogeneous processor architectures. *IEEE Transactions on Parallel and Distributed Systems*, 4, 175–187.
32. S. Chakravarty, C. Ravikumar, and D. Choudhuri. 2002. An evolutionary scheme for co-synthesis of real-time systems. In *Proceedings of Design Automation Conference*, pp. 251–256.
33. C. Lee and S. Ha. 2005. Hardware–software co-synthesis of multitask MPSoCs with real-time constraints. In *Proceedings of Sixth International Conference on ASIC*, pp. 919–924.

34. K. Ramamritham and J. A. Stankovic. 1990. Efficient scheduling algorithms for real-time multiprocessor systems. *IEEE Transactions on Parallel and Distributed Systems*, 1, 184–194.
35. H. Oh and S. Ha. 2002. Hardware–software co-synthesis of multimode–multitask embedded systems with real-time constraints. In *Proceedings of Tenth International Symposium on Hardware–Software Codesign*, pp. 133–138.
36. Y. Kim and T. Kim. 2006. HW/SW partitioning techniques for multimode–multitask embedded applications. In *Proceedings of 16th ACM Great Lakes Symposium on VLSI*, pp. 25–30.
37. A.N. Audsley, A. Burns, M. Richardson et al. 1993. Applying new scheduling theory to static priority pre-emptive scheduling. *Software Engineering Journal*, 8, 284–292.
38. C.L. Liu and J.W. Layland. 1973. Scheduling algorithms for multiprogramming in a hard-real-time environment. *Journal of ACM*, 20, 46–61.
39. A. Awwal. 2008. Co-synthesis of Multiple Processor Embedded Systems for Real-time Applications, MASc. Thesis, Ryerson University, Toronto Canada.
40. G.N. Khan and U. Ahmed. 2008. CAD tool for hardware–software co-synthesis of heterogeneous multiple processor embedded architectures. *Design Automation for Embedded Systems*, 12, 313–343.
41. T.Y. Yen and W.H. Wolf. 1995. Communication synthesis for distributed embedded systems. In *Proceedings of International Conference on Computer-Aided Design*, pp. 288–294.
42. E3S: Embedded System Synthesis Benchmarks Suite. http://ziyang.eecs.umich.edu/~dickrp/e3s/

3 Methods for Non-Intrusive Dynamic Application Profiling and Soft Error Detection

Roman Lysecky and Karthik Shankar

CONTENTS

3.1 INTRODUCTION

Exponential advances in process technologies and circuit capacities led to the current success and future promise of high-performance computing enabled by multicore principles. However, advances in integrated circuits present several key challenges to system reliability. Transient and permanent errors such as electromigration, negative bias temperature instability (NBTI), time dependent dielectric breakdown (TDDB), and thermal cycling (TC), are expected to increase [14,29]. To be successful, computing systems must be able to continue functioning in spite of these soft errors, necessitating the development of new methods for self-healing circuits that can detect and recover from these errors.

Within multicore computing systems, soft errors can affect both the functionality and performance of software executing on each processor. Because software can be executed on any processor core, multicore systems offer a significant advantage over single core alternatives. If a soft error occurs within a specific core—transient or permanent—the software tasks executing on that core can be relocated to another processor core. Such a self-healing system should monitor the executing software tasks to identify soft errors and recover from them by re-executing the affected software tasks or migrating the affected tasks to another processor core. To check that the transient error has passed and clear the core of any permanent errors, it is necessary to re-evaluate the error states of previously affected cores to determine whether they can be safely utilized again and under what conditions.

A self-healing system is capable of monitoring its own execution to isolate the sources of soft errors and recover from them by reconfiguring the system to perform the same computation. For multicore software implementations, it is minimally necessary to detect soft errors that affect the correctness and/or performance of software execution and recover from these faults by executing the affected software regions using other available processing resources.

Most existing approaches for detecting soft errors within the execution of a software application are intrusive and require insertion of additional instructions into the software binary or requiring additional hardware components within the microprocessor. For embedded systems, such methods potentially change the behavior of an application and incur significant runtime overhead.

In the cases of real-time systems that are usually designed with tight timing constraints, the slightest runtime overhead can lead to missed deadlines and potential system failure [5]. Alternatively, most hardware-based approaches directly incorporate redundant computational units or specialized components for detecting errors that can affect the critical path of a microprocessor. In addition, these approaches assume a system developer has access to the microprocessor design. For many embedded systems, processor cores are provided as hard macros or encrypted sources, thus limiting developers' ability to incorporate hardware-based soft error detection methods directly within a processor.

In this chapter, we provide an overview of non-intrusive dynamic application profiling and demonstrate how such runtime profiling can be effectively adapted to detect soft errors at the application level. We first present a dynamic application soft error detector (DASED) capable of non-intrusively detecting soft errors during the execution of a software application without requiring modification to the software application or microprocessor hardware. The DASED design utilizes non-intrusive profiling methodology to monitor the control flow of a software application. This statically determined profiling is based on known execution characteristics of an application's tasks. We further present an area-efficient implementation, referred to as the control focused soft error detector (CNFSED), capable of providing the same level of soft detection with a small area footprint by removing and/or redefining the underlying profiling method specifically for soft error detection. The DASED and CNFSED designs both achieve error detection rates above 90% for control errors and 85% of unmasked errors while incurring minimal area overhead.

3.2   RELATED WORK

A popular method for detecting and recovering from errors is fault-tolerant computing. This type of computing utilizes redundancies (e.g., triple modular redundancy) [27] within the computing resources to reduce the chance that a soft error will affect the final outcome of an application or circuit. A fault-tolerant software system may utilize multiple processor cores that all execute the same task. The results from the redundant executions are compared to determine whether a consensus of the final result can be reached. For example, if three or more of five redundant computations match, the result is likely correct.

By utilizing redundancy, the probability of an error affecting more than one processor is significantly reduced and the system can continue to execute correctly as long as a specified number of processors produce the same result. While effective, fault-tolerant computing requires significant duplication of resources as redundancy is required at every level of the computation, i.e., a fault-tolerant computing system needs redundant processors, caches, and memories. In addition, concurrent execution of redundant computations increases power consumption in proportion to the level of redundancy. The additional circuitry required to compare the outputs of the redundant computations may also affect the system performance. As cost and power consumption are critical design constraints, fault-tolerant computing is prohibitive for many users.

DIVA [3] is a dynamic error detection and recovery technique with integrated functional redundancy within the retirement stage of an out-of-order microprocessor. The *checker* components incorporated within the retirement stage verify the correctness of all computations and communication operations re-executing the retired instructions in order. If a computation or memory address was incorrectly computed, the entire pipeline of the processor will be flushed and the errant instruction re-executed.

This approach requires that the DIVA checker components be functionally and electrically reliable to avoid errors within the components. By re-executing all computations, DIVA is capable of detecting and recovering from most soft errors, but comes at the unavoidable tradeoff of significantly increased area requirements and performance overhead.

Argus [18] utilizes a combination of computational redundancy, parity bits within the memory subsystems, and dynamic data flow verification [19] to verify both the control and data flow of software execution for simple processor cores. To verify the control and data flow of an application, Argus inserts signatures within the basic blocks of the application to indicate the set of valid successor blocks. At runtime, the control and data flow checkers can then verify the correct execution sequence by verifying that each subsequently executed basic block adheres to the statically determined execution order. Overall, Argus achieves a 98% error detection rate for unmasked errors and requires only 10% area overhead. However, it incurs a performance penalty of 4% on average that may reach as high as 19%.

Significant research in soft error detection [8,11,13,16,22,24] focused on utilizing statistically determined application execution behaviors to detect anomalous activities at the application or system level. Many of these approaches utilize data value profiling to determine the expected values of data stored to memory. For example, by comparing the actual values stored by a particular instruction at runtime to a statically expected range of values obtained using training data sets, the onset of a soft error can be detected if the actual data value lies outside the expected range.

The Software Anomaly Treatment (SWAT) project [11,16,24] utilizes a combination of lightweight firmware and hardware buffers to detect anomalous application execution behavior by monitoring and verifying likely program invariants that define expected data values for store operations. SWAT is capable of detecting over 95% of soft errors with a performance overhead of 5 to 14% depending on the target processor.

Proposed software-based approaches to detecting soft errors insert duplicate instructions within the compiled software application [8,13]. Hu et al. [13] proposed an approach in which the output of the original instruction execution is stored in a hardware queue. When the duplicate instruction is executed, the result can be compared to the original value. Interestingly, this approach allows designers to trade off error detection rate with the resulting performance and energy overhead by controlling the number of duplicated instructions.

Shoestring [8] utilizes compiler support to select a small subset of instructions that are most likely to result in perceivable errors at the application level. Results demonstrated that Shoestring is capable of detecting 80% of unmasked errors with an average performance penalty of only 16%.

Other approaches [15,28] originally targeted at detecting malicious code execution can be utilized to detect soft errors that affect the control flow of a software application. These approaches rely on verifying the function call execution sequence or verifying the target address for branch instructions by tracking which instruction addresses are valid sources for reaching a particular instruction destination.

BISER [20] is a hardware-based soft error detection and recovery approach that utilizes redundant latches and specialized hardware inserted within the processor design to both detect and correct soft errors at runtime. Soft errors can be detected by identifying the point at which the data contained in the main and redundant latches differ. When errors are detected, specialized hardware can be utilized to return to the previously valid value in order to recover from the error. This detection and recovery capability can be accomplished within only a 5% performance overhead.

Although not intended for detecting errors, Razor [7] monitors the error rate of a system at runtime to reduce the overall power consumption. For flip-flops along the critical path, Razor incorporates shadow latches connected to a delayed clock. Timing errors can be detected at runtime by observing differences in the main flip-flop and shadow latch. To reduce power consumption, Razor dynamically reduces the operating voltage of the processor until a certain threshold error rate is observed, under the assumption that that the performance penalty of re-executing software code due to the timing errors is acceptable given the resulting power reduction.

Overall, the various software and hardware-based approaches for soft error detection achieve excellent detection rates, ranging from 80 to 98% of unmasked errors. However, while the associated performance overheads of 4 to 16% may be tolerable for many systems, we seek to develop a *non-intrusive* detection method targeted at embedded and real-time systems that requires no modifications to the application or microprocessor and incurs no performance overheads.

3.3 OVERVIEW OF NON-INTRUSIVE DYNAMIC APPLICATION PROFILING

Application profiling—the process of monitoring an application to determine the frequency of execution within specific regions—is an essential step of the design process for many software and hardware systems. Profiling has long been utilized to identify the most frequently executed regions of a software application, and software developers can focus their efforts on optimizing those regions. Most previous profiling approaches intended for desktop computing introduce runtime overhead by inserting additional code into an application or interrupting the processor at specific intervals to sample the registers.

A common software-based profiling approach involves application instrumentation by adding code to count frequencies of the desired code regions [9,21]. For example, if we wish to count the frequency of execution of a subroutine, we can add code to the beginning of the subroutine that increments an associated counter variable. To reduce runtime overhead, other profiling approaches use statistical sampling techniques [1,6,32] that interrupt the microprocessor at certain intervals or create an additional software task for profiling, and then read the program counter and other internal registers to determine execution behavior statistically.

For embedded systems, both instrumentation and statistical profiling approaches potentially change the behavior of an application and incur significant runtime overhead. In the case of real-time systems that are usually designed with very tight timing constraints, the slightest runtime overhead can lead to missed deadlines and potential system failure. For further details and discussion of existing software- and hardware-based profiling techniques, we refer readers to [30].

We provide an overview of a non-intrusive dynamic application profiler (DAProf) [21,25] capable of profiling an executing application by monitoring its short backward branches, function calls, and function returns, as well as efficiently detecting context switches to provide accurate characterization of frequently executed loops within multitasked applications.

Figure 3.1 presents an overview of DAProf and highlights its integration within a microprocessor-based system. DAProf non-intrusively monitors the microprocessor's instruction bus to detect the addresses of short backward branches, function calls and function returns, and context switches. While the DAProf can decode the instructions on the instruction bus, we currently assume the microprocessor provides a one-bit output; *sbb* indicates a short backward branch has been executed; a one-bit output *func* signal indicates execution of a function call; and a one-bit output *ret* indicates a function has returned.

Figure 3.2 presents the internal architecture of the DAProf design with support for multitasked applications. DAProf consists of a profiler task filter for specifying the tasks to filter and detecting context switches, a profiler FIFO to synchronize the microprocessor and profile cache, a profile cache that stores all relevant profile statistics for loops being profiled, and a profiler controller for analyzing short backward branches, function calls, function returns, and context switches to update the profiling statistics within the profile cache.

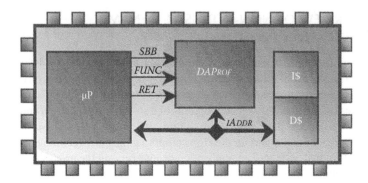

FIGURE 3.1 Overview of Dynamic Application Profiler (DAProf) integration with microprocessor system utilizing signals for detecting short backward branches, function calls, and function returns.

The *Profiler Task Filter* is primarily utilized to non-intrusively detect context switches between the tasks being profiled. It is implemented as a designer programmable array storing the starting and ending address of each task or any region of code to be profiled. The filter provides great flexibility in profiling a multitasked application by allowing a designer to selectively profile specific tasks, functions, library code, systems calls, etc., while filtering out elements that are not of immediate interest. To detect context switches, the task filter monitors the processor's instruction bus to determine which task is currently executing or that no profiled tasks are executing. The device also filters the inputs from the processor for all non-profiled regions of code. In other words, if the currently executing instruction does not fall within a profiled task or code region, the *sbb*, *func*, and *ret* inputs from the processor will be ignored.

The *Profiler FIFO* monitors the *sbb*, *func*, *ret*, and *cs* signals from the task filter. Whenever a profile event is detected, the FIFO stores the *Tag* and *Offset* for short backward branches, the originating addresses of function calls, the return addresses of function returns, or the instruction address immediately after a context switch provided by the task filter. Internally, the FIFO also stores an encoding indicating whether an entry is a short backward branch, function call, function return, or context switch.

In addition, the FIFO is used to synchronize the operating frequency of the microprocessor and task filter and the operating frequency of the internal DAProf design because the microprocessor and task filter may operate at a higher clock frequency. As short backward branches do not occur on every clock cycle, the internal design need not operate at the same frequency of the microprocessor. A typical loop of interest within a software application consists of at least two instructions in addition to the short backward branch at the end of the loop. We experimentally determined that the smallest profiled loop within the applications considered consisted of four instructions.

It is sufficient to assume that short backward branches on average occur no more than once every four instructions, indicating that the internal design can efficiently operate at one-fourth the operating frequency of the microprocessor. However, the FIFO should be large enough to accommodate bursts of short backward branch activity that may occur periodically as an application executes.

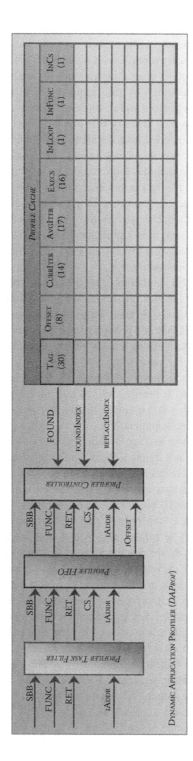

FIGURE 3.2 Architectural overview of Dynamic Application Profiler (DAProf) supporting multitasked applications consisting of a profiler task filter, FIFO, controller, and profile cache (bit widths for profile cache entries are in parentheses).

In addition, DAProf needs to monitor function calls, function returns, and context switches. The combined frequency of all events is not expected to increase the maximum expected frequency of such events. This is evident in the fact that both function calls and function returns require several instructions for maintaining the application's execution stack that limits their overall frequency. Similarly, context switches require dozens of instructions to store and restore task contexts.

DAProf's profile cache is small memory that maintains current profiling results and intermediate information needed for loop identification, iteration and execution profiling statistics, and loop execution monitoring. The main profiling information stored in the profile cache includes loop executions, average iterations per loop execution, and loop iterations for the current execution. Profiled loops are identified within the profile cache by the address of the loop's short backward branch, which serves as the *Tag* entry for the cache, and by the loop's *Offset*. The profile cache's *Tag* is a 30-bit entry that stores the most significant 30 bits of a loop's short backward branch address. The profile cache's *Offset* is an 8-bit entry that corresponds to the size of the loop in number of instructions.

Loop Executions provide the number of times a loop has been executed throughout the application execution. DAProf utilizes a 16-bit entry for loop executions that allows 65,536 loop executions to be profiled without saturations. Whenever a loop's executions become saturated, DAProf's profiler controller will adjust the loop executions for all entries, thereby maintaining a list of relative executions and ensuring all entries do not eventually become saturated.

Current Iterations provides the number of times a loop has iterated for the current loop execution and is stored in the profile cache as a 14-bit entry that allows DAProf to accurately profile loops with a maximum of 16,384 iterations per execution, which is well suited for most embedded applications.

Average Iterations stores the average number of times a loop iterates per loop execution. As many loops do not iterate a fixed number of times per execution, the average iterations cannot be accurately stored as an integer value. Instead, the profile cache stores the average iterations as a 17-bit fixed point number using 14 bits to represent the integer portion and 3 bits for the fractional part.

The profile cache contains a 1-bit *InLoop* flag that is utilized to indicate a loop is currently being executed. The *InLoop* flag is essential in determining whether the execution of a short backward branch corresponds to a new loop execution or an additional iteration for the current execution. A 1-bit *InFunc* flag is utilized to indicate a loop has called a function that is currently being executed. In addition, a 1-bit *InCS* flag is utilized to indicate a loop's execution has been interrupted due to a context switch. The *InFunc* and *InCS* flags are essential in ensuring that the *InLoop* flag for a loop that has called a function or whose execution has been interrupted due to a context switch is not reset incorrectly.

The *Associativity* of the profile cache potentially provides tradeoff between cache size or performance and profiling accuracy. With a fully associative profile cache, the replacement policy must compare all entries within the cache to determine the entry with the fewest total iterations, thereby requiring large hardware resources and reducing the overall performance of the DAProf. While, a fully associative profile cache may provide better profiling accuracy as the replacement policy can select among all cache entries, decreasing the associativity of the profile cache provides increased performance and smaller area requirements by reducing the number of entries the replacement policy must consider.

The replacement policy incorporated within the profile cache uses total loop iterations to determine which entry will be replaced when a new loop is executed, where the entry with the fewest total iterations will be replaced. The total loop iterations are calculated as the product of average iterations and executions. While this policy performs relatively well on its own, newly executed loops may not execute or iterate quickly enough to avoid being replaced immediately.

To address this, the profile cache includes a 3-bit loop *Freshness* value that represents how recently a loop has been executed or iterated. A larger freshness indicates a loop has been executed more recently. The freshness value is utilized within the replacement policy to consider for replacement only loops that are not fresh. A loop that is not fresh has a freshness value of zero. A 3-bit freshness entry allows up to seven loops per task to be considered fresh and allows newly executed loops to be profiled for an extended duration before their profile cache entries are considered during replacement.

The profiler controller interfaces with the profiler FIFO and updates the profiling results for the current loops within the profile cache. The profiler controller either receives the *sbb* signal along with the calculated branch offset, *iOffset*, *func*, *ret*, or *cs* signal from the profiler FIFO in addition to *found*, *foundIndex*, and *replaceIndex* signals from the profile cache. The *found* and *foundIndex* signals indicate whether the current short backward branch is found in the profile cache and at what location. The *replaceIndex* provides the index for the loop entry that will be replaced if the current short backward branch is not found. In all cases, the address of the instruction of interest is provided by the *iAddr* signal from the profiler FIFO.

Whenever a short backward branch is detected, the profiler controller will determine whether the loop is found within the cache. If the loop is found and the loop is currently executing—as indicated by the loop's *InLoop* flag—the short backward branch execution indicates a loop iteration has been detected and its current iterations are incremented. Otherwise, if the loop is not currently being executed, a new loop execution is detected. For new loop executions, the profile controller increments the loop's executions, sets the *InLoop* flag, sets the current iterations to one, decrements the freshness value for all other loops within the current task, and sets the freshness of the current loop to the maximum.

Finally, if the profiler controller detects that the loop's executions have become saturated, the executions for all loops will be divided by two. In addition to ensuring that the executions for all loops never become saturated, this approach provides a mechanism for monitoring the dynamic nature of an application in which loops that were considered important may no longer be executed as time progresses. Initially, a previously executed loop's high number of total iterations may ensure the loop is not replaced during profiling. However, after several saturations are encountered, the reported total iterations will be decreased relative to other loops and can be replaced if a loop is no longer executed.

If a loop's backward branch is not found within the profile cache, the profiler controller will replace the entry within the cache as indicated by *replaceIndex*.

The profiler controller initializes this profile cache entry by setting the *Tag* and *Offset* to those of the newly profiled loops, setting the executions to one, setting the *InLoop* flag, setting the current iterations to one, decrementing the freshness value for all other loops within the current task, and setting the freshness of the newly executed loop to the maximum.

Whenever a context switch is detected, the profiler controller first sets the *InCS* flags for all currently executing loops, i.e., loops whose *InLoop* flags are still set. This can be efficiently implemented simply by copying all *InLoop* entries to the corresponding *InCS* entries in the profile cache. The profiler controller then determines which loops, if any, will resume execution as the result of the context switch. Thus, if an address after a context switch falls within the bounds of any loops whose *InLoop* flags are set, the *InCS* flags for those loops are reset.

Whenever a function call is detected, the profiler controller sets the *InFunc* flags for all currently executing loops, i.e., those whose *InLoop* flags are still set. This can be efficiently implemented simply by copying all *InLoop* entries to the corresponding *InFunc* entries within the profile cache. Whenever a function return is detected, the profiler controller resets the *InFunc* and *InCS* flags for loops that contain the address of the function return's destination, i.e., the loops from which the corresponding function was called. If a function call is executed from the innermost loop of a nested loop, the *InFunc* flags for all loops within the nested structure will be set. On return from that function call, the profiler controller must reset the *InFunc* flags for all loops within the nested loops.

For all profiling events, the profiler controller checks all entries of the profile cache whose *InLoop* flags are set to determine whether the application is still executing within those loops. The profiler controller also utilizes the *InFunc* and *InCS* flags to ensure that the *InLoop* flag is not incorrectly reset during a function call or context switch. For all detected short backward branches, function calls, function returns, and context switches, the profiler controller checks all entries of the profile cache whose *InLoop* flag is set and whose *InFunc* and *InCS* flags are not set to determine whether the application is still executing within those loops. If a loop is no longer executed, the profile controller resets the *InLoop* flag and updates the loop's average iterations.

DAProf is capable of non-intrusively profiling an executing application by monitoring the application's short backward branches, function calls, function returns, and context switches to provide detailed execution statistics for software-critical loops. By utilizing efficient dedicated hardware circuitry to monitor the instruction bus of a target processor, DAProf can profile multiple tasks of a software application with 98.5% and 99.5% accuracy for average iterations and loop executions, respectively, using only 10% additional area compared to an ARM9 processor. Importantly, this profiling methodology serves as the basis for developing efficient non-intrusive soft error detection methods discussed in the following sections.

3.4 DYNAMIC APPLICATION SOFT ERROR DETECTION

Figure 3.3(a) presents an overview of DASED's integration within a microprocessor system. DASED non-intrusively monitors a microprocessor's trace/debug port to detect short backward branches, function calls, function returns, and context

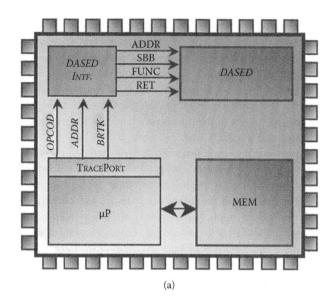

(a)

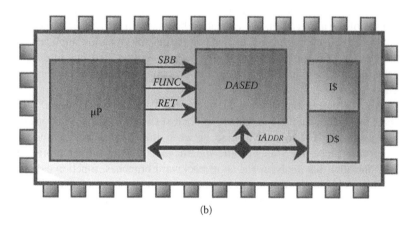

(b)

FIGURE 3.3 Overview of Dynamic Application Soft Error Detector integration with microprocessor system (a) utilizing signal processor's trace port required by DASED design or (b) directly monitoring microprocessor instruction bus to detect short backward branches.

switches. It considers a short backward branch as any branch instruction whose target address is a negative offset below 1024, which corresponds to small loops containing fewer than 256 instructions.

The DASED design requires as input a 1-bit signal *sbb*, indicating a short backward branch has been executed, a 1-bit *func* signal indicating a function call is being executed, and a 1-bit *ret* signal indicating a function has returned. To detect function calls and returns, DASED requires the address from which a function was called and the address to which it returned. As these exact signals are often not available directly from the target processor, a DASED interface component is utilized to interface to the available signals from the trace/debug port provided by most microprocessors [2,31]. Minimally, the DASED *Interface* component requires the address and opcode for all executed instructions and a 1-bit signal indicating a branch has been taken. Alternatively, the DASED design may detect the required signals by directly monitoring the instruction bus of the processor and decoding instruction as they are executed, as illustrated in Figure 3.3(b).

In contrast to previous application-level soft error detection methods that utilize data value profiling, we focus on loop-level profiling because embedded applications are often loop-centric. Based on the initial software application, an initial profiling or static analysis determines the execution statistics for most critical (frequently executed) loops within an application. These statistics include the instruction address identifying the short backward branch *(Tag)*, the size of the loop *(Offset)*, the minimum number of loop iterations per execution *(MinIter)*, and the maximum number of loop iterations per execution *(MaxIter)*. This initial profiling or analysis step may be executed statically at compile time via static analysis or simulation or dynamically at runtime during an initial execution phase in which the absence of errors can be guaranteed or verified.

During application execution, DASED will dynamically detect errors in the control flow of the application, i.e., execution behavior directly related to branches, function calls, and context switches, by comparing the dynamic application execution behavior with statically determined loop execution statistics to detect soft errors that ultimately manifest in changes in execution behavior of the loops being profiled.

Figure 3.4 presents the internal architecture of the DASED design. DASED consists of a profiler task filter for detecting context switches and instruction address errors, a profiler FIFO to synchronize between the microprocessor and profile cache, a profile cache that stores profile statistics for loops being monitored, and a DASED controller that analyzes the short backward branches, function calls, function returns, and context switches to monitor dynamic execution behavior of loops to detect control errors. As an extension to the DAProf design, we explain the additional functionality supported by DASED for soft error detection in the following sections.

3.4.1 Profiler Task Filter

The profiler task filter is used primarily to non-intrusively detect context switches among the tasks being monitored. In the DASED design, the task filter also monitors the instruction addresses to detect errors related to the execution of instructions that are incorrectly aligned or do not correspond to valid instruction addresses in the application's address space. In addition to storing the address bounds for each task to be monitored, the task filter also stores the entire range of valid addresses for all codes within the application including library and operating system codes. If the instruction address does not fall within the valid address bounds or is incorrectly aligned, the task filter will directly assert an error.

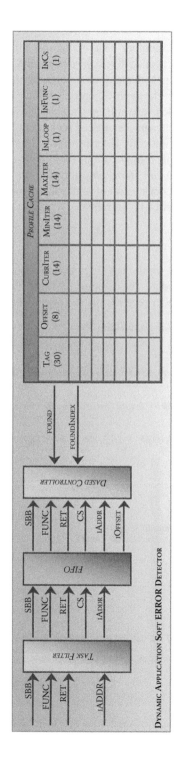

FIGURE 3.4 Architectural overview of Dynamic Application Soft Error Detector (DASED) consisting of task filter, FIFO, controller, and profile cache (bit widths for profile cache entries are in parentheses).

3.4.2 PROFILE CACHE

The profile cache is a small memory that maintains current profiling results, intermediate state information needed for loop identification and monitoring, and statically determined minimum and maximum iteration statistics for each loop. Again, we consider a profile cache with 32 entries, which is sufficiently large to profile and monitor the embedded software applications considered herein.

In addition to the profile cache entries utilized within DAProf, DASED incorporates profiling information for statically determined minimum and maximum loop iterations along with the dynamic loop iterations for a loop's current execution. The *Minimum Iterations* and *Maximum Iterations* store the minimum and maximum number of times a loop iterates per execution in the absence of errors, and store the times within the profile cache as 14-bit entries.

3.4.3 CONTROLLER

Figure 3.5 presents pseudocode for the DASED controller operation. The controller interfaces with the FIFO and updates the dynamic profiling information for the loops in the profile cache. The controller receives the *sbb* signal along with the calculated

DASED (*iAddr, iOffset, sbb, func, ret, cs, found, foundIndex*):

```
1.  if ( cs ) {
2.      for all i, InCS[i] = InLoop[i]
3.      for all i, if ( InLoop[i] && (iAddr <= Tag[i] && iAddr >= Tag[i]-Offset[i] ) InCS[i] = 0
4.  }
5.  if ( func ) {
6.      for all i, if ( iAddr <= Tag[i] && iAddr >= Tag[i]-Offset[i]) InFunc[i] = InLoop[i]
7.  }
8.  else if ( ret )
9.      for all i, if ( (InFunc[i] || InCS[i]) && (iAddr <= Tag[i] && iAddr >= Tag[i]-Offset[i]) ) {
10.         InFunc[i] = 0
11.         InCS[i] = 0
12.     }
13. }
14. else if ( sbb ) {
15.     if ( found ) {
16.         if ( InLoop[foundIndex] ) {
17.             CurrIter[foundIndex] = CurrIter[foundIndex] + 1
18.             if ( CurrIter[foundIndex] > MaxIter[foundIndex] ) return ERROR
19.         }
20.         else {
21.             CurrIter[foundIndex]    = 1
22.             InLoop[foundIndex]      = 1
23.         }
24.     }
25. }
26. if ( sbb || func || cs ) {
27.     for all i, if ( InLoop[i] && !InFunc[i] && !InCS[i] && !(iAddr <= Tag[i] && iAddr >= Tag[i]-Offset[i]) ) {
28.         InLoop[i] = 0
29.         if ( CurrIter[foundIndex] < MinIter[foundIndex] || CurrIter[foundIndex] > MaxIter[foundIndex] )
30.             return ERROR
31.     }
32. }
```

FIGURE 3.5 Pseudocode for Dynamic Application Soft Error Detector controller.

iOffset, or one of *func*, *ret*, or *cs* signals from the FIFO in addition to *found* and *foundIndex* signals from the profile cache. The *found* and *foundIndex* signals indicate whether the current short backward branch is found in the profile cache and at what location. In all cases, the address of the instruction of interest is provided by the *iAddr* signal from the FIFO.

Whenever a short backward branch is detected, the DASED controller will determine whether the loop is found within the cache. If the loop is found and the loop is currently executing (as indicated by the *InLoop* flag), the short backward branch execution indicates detection of a loop iteration and the loop's current iterations are incremented. Otherwise, if the loop is not currently being executed, a new loop execution is detected. For new loop executions, the controller sets the *InLoop* flag and sets the current iterations to one.

Whenever a context switch is detected, the controller first sets the *InCS* flags for all currently executing loops (those whose *InLoop* flags are still set). This can be efficiently implemented simply by copying all *InLoop* entries to the corresponding *InCS* entries within the profile cache. The controller then determines which loops, if any, will resume execution as the result of the context switch. Thus, if the address after a context switch falls within the bounds of any loops whose *InLoop* flags are set, the *InCS* flags for those loops are reset.

Whenever a function call is detected, the controller sets the *InFunc* flags for all currently executing loops (those whose *InLoop* flags are still set). This can be implemented simply by copying all *InLoop* entries to the corresponding *InFunc* entries within the profile cache. Whenever a function return is detected, the DASED controller resets the *InFunc* and *InCS* flags for loops that contain the address of the function return's destination (loops from which the corresponding function was called). If a function call is executed from the innermost loop of a nested loop, the *InFunc* flags for all loops in the nested structure will be set. On return from the function call, the controller must reset the *InFunc* flags for all loops within the nested loops.

For all profiling events, the controller checks all entries of the profile cache whose *InLoop* flag is set to determine whether the application is still executing within those loops. The controller also utilizes the *InFunc* and *InCS* flags to ensure that the *InLoop* flag is not incorrectly reset during a function call or context switch. For all detected short backward branches, function calls, function returns, and context switches, the DASED controller checks all entries of the profile cache whose *InLoop* flag is set and whose *InFunc* and *InCS* flags are not set to determine whether the application is still executing within those loops. If a loop is no longer being executed, the DASED controller resets the *InLoop* flag.

The DASED controller detects control errors by monitoring the loop iterations per loop execution to verify that the current iterations for past loop executions are within the statically determined minimum and maximum iteration bounds. As soon as the current loop iterations exceed the maximum, the DASED controller will assert an error. Additionally, when the end of a loop execution is detected, the current and minimum iterations are compared and an error asserted if the current iterations are fewer than the minimum.

3.4.4 HARDWARE IMPLEMENTATION

The DASED design was implemented in Verilog and synthesized using Synopsys Design Compiler targeting both UMC 180-nm and TSMC 90-nm technologies. With the 180-nm technology, the DASED design required 0.78 mm² with a maximum operating frequency of 843 MHz. The area required for the DASED design is approximately 6.7% of the area of an ARM 9 with a 32-KB cache implemented within the same 180-nm technology. Using the TSMC 90-nm device, the DASED design required 0.14 mm² with a maximum operating frequency of 1.52 GHz. The area required for the 90-nm DASED design is approximately 6% of the area of ARM 1156T implemented within the same technology.

Overall the DASED design can achieve operating frequencies suitable for high-end embedded systems such as the Marvell Kirkwood processor with operating frequencies exceeding 1 GHz with small area overhead. We reiterate that the DASED design is non-intrusive and does not affect the critical path of the target microprocessor or execution of the application monitored. We further note that DASED's profile cache is currently implemented using registers. By re-implementing the profile cache using SRAM, we anticipate that a more area-efficient design can be created in the future.

We further ported our DASED design to a MicroBlaze processor implemented in a Xilinx Virtex-6 FPGA (XC6VLX75T). We developed a DASED interface that utilizes available signals from the MicroBlaze trace port. Specifically, the DASED *Interface* utilizes *Trace_Valid_Instr*, *Trace_Instruction*, and *Trace_PC* signals from MicroBlaze [31]. *Trace_PC* provides the address of the instruction to be executed in the next cycle, *Trace_Instruction* provides the opcode of the current instruction, and *Trace_Valid_Instr* indicates that the *Trace_PC* and *Trace_Instruction* outputs are valid for the current cycle.

The DASED *Interface* monitors the execution addresses and decodes the opcodes for all valid instructions to detect branches, function calls, and function returns to generate the *iAddr*, *sbb*, *func* and *ret* signals needed. We synthesized both the DASED design and interface for Virtex-6 FPGA using Xilinx ISE 11.4. The DASED interface requires only 131 LUTs and 162 flip-flops with a maximum operating frequency of 298 MHz. The complete DASED design requires 9,404 slices and 9,492 flip-flops, roughly corresponding to 20% of available logic resources. The complete DASED design can operate at a maximum clock frequency of 253MHz—well above the operating frequency of the MicroBlaze processor.

3.5 AREA-EFFICIENT OPTIMIZATION FOR DYNAMIC APPLICATION SOFT ERROR DETECTION

As the DASED design was built upon the underlying DAProf profiling framework, the resulting implementation supports both dynamic application profiling and soft error detection. This design can be utilized for efficiently profiling an application during development and detecting soft errors at runtime. Alternatively, by redesigning the underlying the framework to focus only on soft error detection, the area requirements can be significant reduced.

3.5.1 CNFSED: MODIFIED MICROPROCESSOR INTERFACE

Figure 3.6 presents an overview of the CNFSED [26] design's integration in a processor system. CNFSED non-intrusively interfaces with the trace/debug port of a processor to monitor loop and context switch execution behavior of the target multitasked application. Again, loops are identified by short backward branches that correspond to any branch instruction whose target address is a negative offset below 1,024. The CNFSED interface utilizes a subset of available signals from the processor's trace/debug port to provide a 1-bit *brTK* signal indicating a branch being executed was taken, a 1-bit *brNTK* signal indicating a branch being executed is not taken, and an *iAddr* signal indicating the address of the branch instruction.

Based on the initial software application, an initial profiling or static analysis is utilized to determine the execution statistics for the critical (frequently executed) loops within the application. These execution statistics include the instruction address identifying the short backward branch (*Tag*), the context ID of the task to which the loop belongs (*CID*), the minimum number of loop iterations per execution (*Minimum Iterations*), and the maximum number of loop iterations per execution (*Maximum Iterations*). During application execution, CNFSED will detect errors in the control flow of the application by comparing the dynamic execution behavior with the statically determined loop execution statistics to detect soft errors that ultimately manifest as changes in execution behavior of the loops being profiled.

The internal architecture of CNFSED consists of the task filter, FIFO, profile array, and CNFSED controller as shown in Figure 3.7. The task filter is used primarily to detect context switches between the tasks and outputs and the context ID (*CID*)

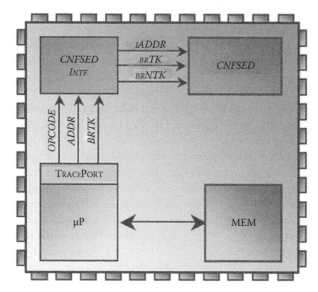

FIGURE 3.6 Overview of Control Focused Soft Error Detector integration with processor system utilizing trace and debug signals commonly available from the processor's trace and debug interface.

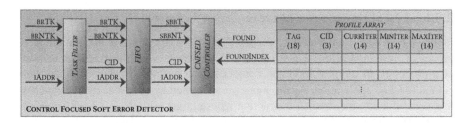

FIGURE 3.7 Architectural overview of Control Focused Soft Error Detection for Embedded Applications (CNFSED) consisting of task filter, FIFO, controller, and profile array (bit widths for profile array entries are in parentheses).

of the currently monitored task non-intrusively. The task filter is implemented as a programmable array storing the starting and ending address of each task or region of code to be monitored.

The task filter maps the address under execution to these bounds to detect the *CID* of the current task. In addition to storing the address bounds for tasks to be monitored, the filter also stores the entire range of valid addresses for all application instructions, including library or operating system code. If the instruction address does not fall within the valid address bounds or aligns incorrectly, the filter will directly assert an error.

The FIFO monitors the *brTK* and *brNTK* signals from the task filter. Whenever a profile event is detected, the FIFO stores the *Tag* for the short backward branch. The FIFO also stores the *CIDs* of the tasks responsible for these signals. It is also used to synchronize the operating frequency of the processor and task filter and the operating frequency of the internal CNFSED design as the processor and task filter may operate at a higher clock frequency.

The profile array is a small memory that maintains current profiling results, intermediate state information needed for loop identification and monitoring, and the statically determined minimum and maximum iteration statistics for each loop. We currently utilize a profile array with 32 entries, which is sufficiently large to monitor the embedded software applications (up to five independent tasks) considered within this chapter, although a larger profile array may be needed for significantly larger applications.

Monitored loops are identified within the profile array by the address of a short backward branch that serves as the *Tag* entry for the array and by the loop's *CID* determined by the task filter. Considering a 32-bit ARM processor and byte address-able memory, the lower two bits for all instruction addresses are identical. The higher bits of the address may be eliminated when *Tag* is used along with the *CID* of the task to identify the loop. We consider the lower 18 bits of the remaining 30-bit address as the *Tag*.

The main profiling information stored in the profile array includes the statically determined 14-bit minimum and maximum loop iterations along with the dynamic loop iterations for current execution. *Current Iterations* provides the number of times a loop has iterated for the current execution and is stored in the profile array as a 14-bit entry.

3.5.2 CONTROLLER

Figure 3.8 presents pseudocode for the CNFSED controller that interfaces with the FIFO and updates the dynamic profiling information for the loops in the profile array. The controller receives the *sbbT* signal along with the *Tag* or the *sbbNT* signal from the FIFO in addition to *found* and *foundIndex* signals from the profile array. The *found* and *foundIndex* signals indicate whether a current short backward branch is found for the current context in the profile array and at what location.

In all cases, the address of the instruction of interest is provided by the *iAddr* signal from the FIFO. Whenever a short backward branch is detected, the CNFSED controller will determine whether the loop is found within the array. If the loop is found, the short backward branch execution indicates detection of a loop iteration and the loop's current iterations are incremented.

Whenever a branch-not-taken signal is detected, the address of the branch within the array is searched. Finding the address indicates that the loop execution is complete. At this time, the *Current Iterations* value can detect any error. The *Current Iterations* value is then reset to zero indicating completion of a loop execution.

The CNFSED controller detects control errors by monitoring the iterations per loop execution to verify that the current iterations for the past loop execution are within the statically determined minimum and maximum bounds. As soon as the current loop iterations exceed the maximum, the CNFSED controller will assert an error. Additionally, when the end of a loop execution is detected, the current and minimum iterations are compared and an error asserted if the current iterations are fewer than the statically determined minimum.

CNFSED (iAddr, sbbT, sbbNT, found, foundIndex):

```
1.  if ( sbbT ) {
2.    if ( found ) {
3.      CurrIter[foundIndex] = CurrIter[foundIndex] + 1
4.      if ( CurrIter[foundIndex] > MaxIter[foundIndex] )
5.        return ERROR
6.    }
7.  }
8.  if ( sbbNT ) {
9.    if ( found ) {
10.     if ( CurrIter[foundIndex] < MinIter[foundIndex] ||
11.         CurrIter[foundIndex] > MaxIter[foundIndex] )
12.       return ERROR
13.     CurrIter[foundIndex] = 0;
14.   }
15. }
```

FIGURE 3.8 Pseudocode for CNFSED controller.

3.5.3 HARDWARE IMPLEMENTATION

The CNFSED design was implemented in Verilog and synthesized using Synopsys Design Compiler targeting a TSMC 90-nm technology. The entire CNFSED design including the interface to the target processor required 0.069 mm^2 with a maximum operating frequency of 1.52 GHz. The CNFSED design required approximately 3% of the area of an ARM 1156T implemented within the same 90-nm technology.

Compared with the DASED design, CNFSED implementation requires 51% less area with equivalent error detection performance. Overall, the CNFSED can achieve operating frequencies suitable for high-end embedded systems while incurring little area overhead. We reiterate that CNFSED is a non-intrusive design that does not affect the critical path of a target processor or execution of an application being monitored.

3.6 EXPERIMENTAL RESULTS

We utilize a taxonomy for soft errors that defines three types: control errors, silent data corruption errors, and masked errors. This taxonomy is specifically related to single event upsets and is useful in categorizing errors and their effects on an application's execution.

Control errors that affect the control flow of an application, leading to noticeable changes in execution or even catastrophic failure, e.g., different instruction executed or fetching instruction from memory outside valid application address bounds. Silent data corruption errors affect only data outputs from an application, e.g., producing incorrect pixel values in a video stream; they do not affect the control flow of an application. Masked errors affect neither the control flow of an application nor the data outputs from the application, e.g., a soft error may affect the value used in a comparison for which the resulting comparison yields the same result. Conversely, unmasked errors are defined as those that affect the control flow or data outputs of an application. In other words, unmasked errors represent the set of all control and silent data corruption errors.

To determine whether an inserted single event upset manifests as a control, silent data corruption, or masked error, we developed an automated simulation environment capable of determining whether an inserted error is a control, silent data corruption, or masked error. We first simulated each application using the SimpleScalar simulator [4] without errors to determine an error-free instruction trace along with tracing all data outputs from the application execution. This error-free trace can then be utilized later to determine whether an inserted single event upset in an application execution manifests as a control, silent data corruption, or masked error.

Applications were simulated again with a single event upset error randomly inserted within the application execution. The inserted errors were modeled as random individual bit flips within the input from or output to all registered components in the processor including all registers within the register file, program counter, status registers, flag registers, pipeline registers, bus addresses, and data inputs and outputs. During a randomly selected instruction cycle, a random registered component was selected and a single random bit flip inserted.

The resulting instruction and data traces for the application execution within soft errors are generated. If the instruction traces do not match, the inserted error is considered a control type as it changes the instruction execution sequence. On the other hand, if only the data outputs from the two executions do not match, the inserted error is considered a silent data corruption. If neither the instruction traces nor data traces differ, the inserted error is considered a masked type.

We performed extensive experiments using our automated simulation and soft error injection environment for 12 single task applications and 14 multitasked applications consisting of 2 to 5 tasks. Each task corresponded to an application from the MiBench benchmark suite [10]. Table 3.1 presents an overview of the single- and multitasked applications considered. Each application is labeled *MTx.y*, where *x* indicates the number of tasks and *y* provides a unique identifier for applications with set numbers of tasks.

Each application was executed 100 times during which a single event upset was randomly inserted. Table 3.2 presents a breakdown of the percentage of silent data corruption errors (*%SDCE*), percentage of control errors (*%CE*), and percentage of masked errors (*%ME*) for each application and highlights the percentage of detected control errors (*%DCE*), percentage of detected unmasked errors (*%DUE*), mean time to detection with 90% confidence interval, and maximum time to detection for all applications considered. Although both DASED and CNFSED can operate as fast as 1.52 GHz, for evaluating the mean and maximum time to detection, we assumed a microprocessor operating frequency of 1 GHz.

On average, both designs could detect 92% of all control errors and 85% of all unmasked errors across the applications considered. However, for a few applications, specifically *MT1.fft*, *MT1.rawcaudio*, and *MT1.rawdaudio*, the unmasked error detection rate was less than 50%. The low detection rate can be attributed partially to a higher prevalence of silent data corruption errors that affect output without affecting application control flow and cannot be detected using our design.

It is important to note that for the application considered, the total percentage of execution time encapsulated by the top 32 executed loops is on average 69% of the execution time. Although we monitored only a small fraction of the execution time, errors within non-monitored regions can still be detected because they are likely to affect the control flows of the monitored regions, thus accounting for the observed overall unmasked error detection rate of 85%. However, as the source of an error may not be monitored directly, increased times to detection may be required.

For errors that are detected, the mean time to detection is on average 9869 ± 1809 μs. The variation in detection time between applications and cross multiple simulations of the same application are results of both the type of error and the detection method (DASED or CNFSED). For example, if a soft error causes invalid execution of an instruction at an address outside the application bounds, the error can be detected within a few cycles, and in many cases at zero cycles. However, if a soft error results in a change in the execution behavior of a loop, the detection time is affected by both loop size and the maximum iteration bound for the loop.

TABLE 3.1
Overview of Single- and Multitasked Applications from MiBench Benchmark Suite

	cjpeg	djpeg	fft	tiff2bw	tiff2rgba	susan	dijkstra	bitcount	stringsearch	qsort	rawcaudio	rawdaudio
MT2.1			—		—							
MT2.2	—									—		
MT2.3						—	—					
MT2.4		—		—								
MT2.5								—	—			
MT2.6	—	—										
MT2.7				—	—							
MT3.1					—			—				—
MT3.2			—	—					—			
MT3.3						—				—	—	
MT3.4				—		—	—					
MT4.1		—				—	—			—		
MT4.2	—	—							—		—	
MT5.1	—		—			—	—			—		

TABLE 3.2

Breakdown of Soft Errors

Application	%CE	%SDCE	%ME	%DCE	%DUE	Time to Detection			
						Mean		Minimum	Maximum
MT1.cjpeg	85	0	15	98	98	4,757	±1106	0	12,880
MT1.djpeg	100	0	0	100	100	6,521	±1039	0	12,909
MT1.fft	66	17	17	61	48	19,606	±7061	0	82,818
MT1.tiff2bw	58	12	30	91	76	30	±34	0	1,080
MT1.tiff2rgba	64	3	33	100	96	181	±114	0	3,073
MT1.susan	34	0	66	100	100	2,135	±846	0	7,899
MT1.dijkstra	64	20	16	88	66	5,391	±2237	0	30,089
MT1.bitcount	74	19	7	80	63	39	±14	0	337
MT1.stringsearch	73	0	27	96	96	265	±139	0	3,100
MT1.qsort	55	0	45	89	89	3,872	±2002	0	37,300
MT1.rawcaudio	76	10	14	41	37	2,552	±535	0	5,416
MT1.rawdaudio	60	30	10	57	40	2,917	±2050	0	25,839
MT2.1	68	7	25	96	87	31,298	±7895	0	127,950
MT2.2	100	0	0	100	100	6,675	±1206	0	16,736
MT2.3	63	1	36	89	88	3,428	±1444	0	26,412
MT2.4	100	0	0	100	100	10,797	±772	0	12,835
MT2.5	75	15	10	97	81	5,024	±1174	0	19,833
MT2.6	100	0	0	100	100	2,866	±818	0	14,409
MT2.7	70	7	23	100	91	335	±101	0	2,026
MT3.1	69	14	17	97	81	3,055	±1114	0	34,954
MT3.3	76	18	6	92	74	48,781	±7283	0	100,988
MT3.3	100	0	0	100	100	13,455	±524	0	15,112

(Continued)

TABLE 3.2 (Continued)
Breakdown of Soft Errors

Application	%CE	%SDCE	%ME	%DCE	%DUE	Time to Detection Mean	Minimum	Maximum
MT3.4	100	0	0	100	100	15,955 ±898	0	18,134
MT4.1	100	0	0	100	100	23,098 ±817	180	24,710
MT4.2	100	0	0	100	100	13,482 ±1528	0	20,560
MT5.1	100	0	0	100	100	30,068 ±4294	0	229,513
Average	**78**	**7**	**15**	**92**	**85**	**9,869 ±1809**	**7**	**34,112**

Percentage of control errors (%CE), silent data corruption errors (%SDCE), and masked errors (%ME) shown for each application, along with percentage of detected control errors (%DCE), detected unmasked errors (%DUE), mean time to detection with 90% confidence interval, and maximum time to detection.

The largest time to detection across all applications was 229,513 μs (229 ms). This implies that while a soft error may have little impact on the application execution at the time of occurrence, its effects may exert lasting long-term impacts on the application execution.

For the applications considered, the soft error detectors did not report any false positives. The loop execution bounds were pessimistically selected for each application using various inputs. However, this does not imply that false positives are not possible with this approach, but rather that the input data sets used did not result in any false positives within our selected loop execution bounds. While tighter statistically determined loop execution bounds may decrease times to detection, a tradeoff of increased occurrence of false positives would be expected.

Additionally, our proposed methodology is sensitive to the execution and statically determined through application profiling, thus requiring sufficiently diverse data inputs to provide good error detection capabilities. We note that further analysis of the tradeoffs among profiling accuracy, varying data inputs, occurrence of false positives, and time to detection is needed and may serve as the focus of future work.

Finally, we further validated the DASED and CNFSED designs by executing several benchmark methods in which errors were directly inserted into software applications using a MicroBlaze processor. For all test cases, both designs correctly detected all inserted errors. We note that these errors were inserted within the application code and not within the MicroBlaze processor, and that a robust framework for dynamically inserting errors into a microprocessor is needed to further evaluate the prototype implementation.

3.7 CONCLUSIONS

The DASED and CNFSED are capable of non-intrusively detecting soft errors within the execution of a software application without requiring modifications to the application or the target processor hardware. For the applications considered, both implementations achieved detection rates of 92% for control errors and 85% for unmasked errors with a mean time to detection shorter than 10 ms. While the DASED design requires only 6 to 7% area overhead and can achieve operating frequencies as high as 1.52 GHz, the CNFSED design requires only 3% area overhead.

REFERENCES

1. Anderson, J., L. Berc, and J. Dean. 1997. Continuous profiling: where have all the cycles gone? *ACM Transactions on Computer Systems*, 15, 4.
2. ARM, Ltd. 2009. CoreSight On-Chip Debug and Real-Time Trace Technology. http://www.arm.com/products/solutions/CoreSight.html
3. Austin, T. 1999. DIVA: a reliable substrate for deep submicron microarchitecture design. International Symposium on Microarchitecture.
4. Burger, D. and T. Austin. 1997. The SimpleScalar Tool Set, Version 2.0. University of Wisconsin Madison Computer Sciences Department Technical Report 1342.
5. Buttazzo, G. 2006. Achieving scalability in real-time systems. *Computer, 39*, 54–59.
6. Dean, J., J. Hicks, C. Waldspurger et al. 1997. ProfileMe: hardware support for instruction-level profiling on out-of-order processors. International Symposium on Microarchitecture.

7. Ernst, D, N. Sung Kim, S. Das et al. 2003. A low-power pipeline based on circuit-level timing speculation. International Symposium on Microarchitecture.

8. Feng, S., S. Gupta, A. Ansari et al. 2010. Shoestring: probabilistic soft error reliability on the cheap. Architectural Support for Programming Languages and Operating Systems.

9. Graham, S.L., P.B. Kessler, and M.K. McKusick. 1982. Gprof: a call graph execution profiler. Symposium on Compiler Construction.

10. Guthaus, M., J. Ringenberg, D. Ernst et al. 2001. MiBench: A free, commercially representative embedded benchmark suite. Workshop on Workload Characterization.

11. Hari, S., M. Li, P. Ramachandran et al. 2009. mSWAT: low-cost hardware fault detection and diagnosis for multicore systems. International Symposium on Microarchitecture.

12. Hazelwood, K. and A. Klauser. 2006. A dynamic binary instrumentation engine for the ARM architecture. Conference on Compilers, Architectures, and Synthesis for Embedded Systems.

13. Hu, J., F. Li, V. Degalahal et al. 2009. Compiler-assisted soft error detection under performance and energy constraints in embedded systems. *ACM Transactions on Embedded Computing Systems*, 8, 27.

14. International Technology Roadmap for Semiconductors. 2008. http://www.itrs.net/

15. Kiriansky, V., D. Bruening, and S. Amarasinghe. 2002. Secure execution via program shepherding. In *Proceedings of USENIX Security Symposium*, pp. 191–206.

16. Li, M., P. Ramachandran, S. Sahoo et al. 2008. Understanding the propagation of hard errors to software and implications for resilient system design. *Computer Architecture News*, 36, 265–276.

17. Malik, A., B. Moyer, and D. Cermak. 2000. A low power unified cache architecture providing power and performance flexibility. In *Proceedings of International Symposium on Low Power Electronics and Design*, pp. 241–243.

18. Meixner, M., M. Bauer, and D. Sorin. 2008. Argus: low-cost, comprehensive error detection in simple cores. *IEEE Micro*, 28, 52–59.

19. Meixner, M. and D. Sorin. 2007. Error detection using dynamic dataflow verification. International Conference on Parallel Architectures and Compilation Techniques.

20. Mitra, S, N. Seifert, M. Zhang et al. 2005. Robust system design with built-in soft error resilience. *Computer*, 38, 43–52.

21. Nair, A. and R. Lysecky. 2008. Non-intrusive dynamic application profiler for detailed loop execution characterization. In *Proceedings of International Conference on Compilers, Architectures, and Synthesis for Embedded Systems*, pp. 23–30.

22. Racunas, P. K. Constantinides, S. Manne et al. 2007. Perturbation-based fault screening. In *Proceedings of International Symposium on High-Performance Computer Architecture*, pp. 169–180.

23. Real-Time Operating System for Multiprocessor Systems. 2008. http://www.rtems.org

24. Sahoo, S., M. Li. P. Ramachandran et al. 2008. Using likely program invariants to detect hardware errors. Conference on Dependable Systems and Networks.

25. Shankar, K. and R. Lysecky. 2009. Non-intrusive dynamic application profiling for multi-tasked applications. In *Proceedings of Design Automation Conference*, pp. 130–135.

26. Shankar, K. and R. Lysecky. 2010. Control-focused soft error detection for embedded applications. *IEEE Embedded Systems Letters*, 2, 127–130.

27. Siewiorek, D. and R. Swarz. 1991. *The Theory and Practice of Reliable System Design*. Bedford, MA: Digital Press.

28. Shi, W., H.H. Lee, L. Falk et al. 2006. An integrated framework for dependable and revivable architectures using multicore processors. In *Proceedings of International Symposium on Computer Architecture*, pp. 102–113.

29. Srinivasan, J. S. Adve, P. Bose et al. 2004. Impact of technology scaling on lifetime reliability. International Conference on Dependable Systems and Networks.

30. Tony J. and M. Khalid. 2008. Profiling tools for FPGA-based embedded systems: survey and quantitative comparison. *Journal of Computers*, 3.
31. Xilinx, Inc. 2008. Setup of a MicroBlaze Processor Design for Off-Chip Trace. http://www.xilinx.com/support/documentation/application_ notes/xapp1029.pdf
32. Zhang, X., Z. Wang, N., Gloy et al. 1997. System support for automatic profiling and optimization. International Symposium on Operating Systems Principles.

4 Embedded Systems Code Optimization and Power Consumption

Mostafa E.A. Ibrahim and Markus Rupp

CONTENTS

4.1 INTRODUCTION

In a growing number of complex heterogeneous embedded systems, the relevance of software components is rapidly increasing. Issues such as development time, flexibility, and reusability are, in fact, better addressed by software-based solutions. Due to the processing regularity of multimedia and DSP applications, statically scheduled devices such as VLIW processors are viable options over dynamically scheduled processors such as state-of-the-art superscalar GPPs.

The programs that run on a particular architecture will significantly affect the energy usage of a processor. The manner in which a program exercises certain parts of a processor will vary the contributions of individual structures to the total energy consumption of the processor. Minimizing power dissipation may be handled by hardware or software optimizations: in hardware through circuit design, and in software through compile-time analysis and code reshaping.

While hardware optimization techniques have been the foci of several studies and are fairly mature, software approaches to optimizing power are relatively new.

Progress in understanding the impacts of traditional compiler optimizations on power consumption and developing new power-aware compiler optimizations are important to the overall energy optimization of systems.

In this chapter, we investigate the impacts of applying high level language optimization techniques on power consumption of embedded systems. Using the example of the Code Composer Studio (CCS) C/C++ compiler, we evaluate the impact of its global performance optimization options -o0 to -o3 on power consumption. Furthermore, we explore the effects of utilizing two architectural features of the targeted processor: namely, the software pipelined loop (SPLOOP) and single instruction multiple data (SIMD) capabilities from a power consumption perspective.

The currently available compiler optimization techniques target either execution time or code size and may improve power consumption only indirectly. These techniques are handicapped for power optimization due to their partial perspectives of the algorithms and their limited modifications to the data structures. Other software optimization techniques like source code transformations can exploit the full knowledge of algorithm characteristics and also modify both data structures and algorithm coding. Interprocedural optimizations are also envisioned. Hence, we investigated several loop, data, and procedural source code transformations from a power consumption perspective.

In Section 4.2, we introduce power-aware optimization methodologies for embedded systems in software. Section 4.3 provides a brief overview of the target architecture along with the experimental setup employed in our experiments. Section 4.4 describes the methodology for measuring power consumption while applying different optimization techniques. Section 4.5 presents the results of investigating global compiler optimization options, specific architectural features, and the impacts of source code transformations on power consumption. Finally, Section 4.6 summarizes the main points of this chapter.

4.2 POWER-AWARE OPTIMIZATION

The reduction of power consumed during the running of software on a microprocessor can be addressed at compiler, low or high level language levels. All the mechanisms present advantages and disadvantages, depending on the target processor and architecture [1]. This section briefly presents the most recent contributions related to optimizing the power consumption of embedded systems from a software perspective.

4.2.1 COMPILER-BASED OPTIMIZATION

Recently, some attempts to investigate the impact of applying standard compiler speed optimization levels to the power and energy consumptions of programmable processors have been introduced. Valluri and John [2] evaluated some general and specific optimizations based on the power and energy consumption of the DEC Alpha 21064 processor while running SpecInt95 and SpecFp95 benchmarks [3].

Ravindran et al. [4] proposed an approach for compiler-directed dynamic placement of instructions into a low-power code cache. They showed that by applying dynamic placement techniques, energy savings may be achieved on the WIMS microcontroller platform. Chakrapani [5] also presented a study of the effect of compiler optimization on energy use by an embedded processor. His work targeted an ARM embedded core, and he used an RTL level model with a Synopsys Power Compiler to estimate power.

Seng et al. [6] reviewed the effects of the Intel compiler general and specific optimizations on energy and power consumption of a Pentium 4 processor running benchmarks extracted from Spec2000 [3]. Zambreno et al. [7] studied power consumption on portable devices. They analyzed the effects of compiler optimizations on memory energy. Based on their results, they concluded that the best optimization approach may not give the best results for power use. Also, they observed that function inlining increased power consumption of the system, unlike loop unrolling that showed a decrease in energy. It is clear that the authors interchangeably used the *power* and *energy* terms.

Casas et al. [8] studied the effects of various compiler optimizations on the energy and power use of the low power C55 DSP of Texas Instruments (TI). Their work did not consider the effects of compiler optimizations on many performance measures such as memory referencing and instructions per cycle that significantly affect power and energy use.

Esakkimuthu et al. [9] compared hardware and software optimizations analyzing cache optimization mechanisms and compiler optimization techniques to lower power consumption. Their results showed that compiler optimizations outperformed cache optimizations in terms of energy savings. Lee et al. [10] investigated compiler transformation techniques to schedule VLIW instructions, aiming to reduce the power consumption of VLIW architectures on the instruction bus. Their experiments showed a noticeable enhancement with four- and eight-way issue architectures for power consumption of the instruction bus as compared to the conventional list scheduling technique.

Kandemir et al. [11] studied the influence of five high level compiler-based optimizations such as loop unrolling and loop fusion on energy consumption using the SimplePower cycle-accurate energy simulator [12], a source-to-source code translator, and a number of benchmark codes. Mehta et al. [13] proposed a compilation technique for lower power consumption based on the energy consumed by the instruction register (IR) and register file decoder.

Leupers [14] analyzed different techniques to design power-efficient compilers and presented a set of software optimization techniques for compiler code generation. Oliver et al. [15] analyzed factors that affected power consumption at the instruction level such as cycles, branches, and instruction reordering. Their experiments demonstrated that software optimization at instruction level is a good approach to minimize energy dissipation in embedded processors.

4.2.2 SOURCE-TO-SOURCE CODE OPTIMIZATION

Most of the software-oriented proposals for power optimization focus on instruction scheduling and code generation, possibly to minimize memory access cost [16]. As expected, standard low level compiler optimizations such as loop unrolling and software pipelining also promote energy reduction since they reduce code execution time. However, a number of cross-related effects cannot be identified clearly and are generally difficult to apply to compilers unless some suitable source-to-source restructuring of the code is applied a priori.

Ortiz et al. [17] investigated the impacts of different code transformations—loop unrolling, function inlining, and variable type declarations—on power consumption. They chose three platforms as targets: 8- and 16-bit microcontrollers and a 32-bit ARM7TDMI processor. Their results show that loop unrolling exerted a significant impact on the consumed power with the 16- and 32-bit processors.

Azeemi et al. [18] examined the effects of loop unrolling factor, grafting depth, and blocking factor on the energy and performance of the Philips Nexperia media processor PNX1302. However, they used the *energy* and *power* terms interchangeably. Hence the improvements in energy obtained from their work are directly related to performance enhancements.

Brandolese et al. [19] stressed the state-of-the-art source-to-source transformations to discover and compare their effectiveness from power and energy perspectives. The data structure, loop and inter-procedural transformations were investigated with the aid of a GCC compiler. The compiled software codes were then simulated with a framework based on the SimpleScaler [20]. The simulation framework was configured with a 1-kByte two-way set-associative unified cache.

Catthoor et al. [21] showed the crucial role of source-to-source code transformations in solving the data transfer and storage bottlenecks in modern processor architectures. They surveyed many transformations aimed at enhancing data locality and reuse.

Kulkarni et al. [22] improved software-controlled cache utilization to achieve lower power requirements for multimedia and signal processing applications. Their methodology took into account many program parameters such as data locality, sizes of data structures, access structures of large array variables, regularity of loop nests, and the size and type of cache. Their objective was to improve cache performance at lower power. The targeted platforms for their research were embedded multimedia and DSP processors.

In the same way, McKinley et al. [23] investigated the impacts of loop transformations on data locality. Yang et al. [24] studied the impacts of loop optimizations on performance and power tradeoffs. They utilized the Delaware Power-Aware Compilation Testbed (Del-PACT), an integrated framework consisting of a modern industry-strength compiler infrastructure and a state-of-the-art micro-architecture level power analysis platform. Low level loop optimizations at the code generation (back end) phase (loop unrolling and software pipelining) and high level loop optimizations at the program analysis and transformation phase (front end, loop permutation and tiling) were studied.

4.3   SOFTWARE AND HARDWARE PLATFORMS

The hardware architecture utilized in this work was the TMS320C6416T fixed-point VLIW DSP from Texas Instruments (TI). This DSP is considered one of the highest performance fixed-point devices [25]. Although the targeted architecture operating

TABLE 4.1

Features of Global Performance Optimization Options

Optimization	Features
-o0	Performs control-flow-graph simplification, allocates variables to registers, performs loop rotation, eliminates unused code, simplifies expressions and statements, expands calls to functions declared inline
-o1	All -o0 optimizations; also performs local copy/constant propagation, removes unused assignments, eliminates local common expressions
-o2	All -o1 optimizations; also performs software pipelining and loop optimizations, eliminates global common sub-expressions and unused assignments, converts array references in loops to incremented pointer form, performs loop unrolling
-o3	All -o2 optimizations; also removes all functions that are never called, simplifies functions with return values that are never used, inline calls to small functions, reorders function declarations so that attributes of called functions are known when caller is optimized, propagates arguments into function bodies when all calls pass same value in same argument position, identifies file-level variable characteristics

frequency ranged from 500 to 1200 MHz, in our set-up, the operating frequency was adjusted to 1000 MHz and the DSP core voltage was 1.2 V.

All measurements were carried out on the TMS320C6416T DSP Starter Kit (DSK) manufactured by Spectrum Digital Inc. The three power test points on this DSK are DSP I/O current, DSP core current, and system current [26]. The current drawn by the DSP core while running an algorithm was captured by the Agilent 34410A 6.5-digit digital multi-meter (DMM). This meter features very high DC basic accuracy (0.003% of the reading plus 0.003% of the range) [27].

The current is captured as a differential voltage drop across a 0.025 Ω sense resistor inserted by the manufacturer in the current path of the DSP core. The input differential voltage drop is divided by 0.025 Ω to determine the current drawn.

The software platform for generating the software binaries to be loaded to the DSP was the embedded C/C++ compiler (Version 6.0.1) in the Code Composer Studio (CCS 3.1) from TI. This compiler accepts C and C++ code meeting the International Organization for Standardization (ISO) standards for these languages and produces assembly language source code for the C6416T device. The compiler supports the 1989 version of the C language.

This compiler features many levels of optimization mainly tuned for speed and/or code size as shown in Table 4.1. A user can invoke the -o0, -o1, -o2, and -o3 options for global speed optimization [28].

4.4 METHODOLOGY AND APPLICATIONS

Let a program X run for T seconds to achieve its goal. V_{CC} is the supply voltage of the system, and I denotes the average current (amperes) drawn from the power source for T seconds. Consequently, T can be rewritten as $T = N \times \tau$ where N is the number of clock cycles and τ is the clock period. The power consumed by running program

X is calculated as $P = V_{CC} \times I$. The amount of energy consumed by X to achieve its goal is given as $E = P \times N \times \tau$ (joules).

Since V_{CC} and τ are fixed for specific hardware, $E \propto I \times N$. However, at the application level, it is more meaningful to consider T instead of N, and therefore energy is expressed as $E \propto I \times T$. This expression shows the main idea in the design of energy-efficient software: reduction of both T and I. From running time (average case) of an algorithm, a measure of T is achieved. However, to compute I, one must consider the average current drawn during execution of a program.

In this chapter, several image and signal processing benchmarks were used to investigate the impacts of invoking different compiler optimization levels on power and energy. The CCS3.1 profiler was employed to profile the benchmarks.

For assessing the effects of SIMD instructions on energy and power consumption, we prepare two versions of the inverse discrete cosine transform (IDCT) algorithm, the discrete cosine transform (DCT) algorithm, and the median filter with a 3×3 window. The first version was implemented without using any SIMD instructions. The second was implemented with all possible SIMD instructions. The functionalities of both versions were tested and verified to yield the same results.

To study the impacts of applying different source code transformations, we first decided the source code transformations to be investigated. Second, we prepared the suitable software kernel to allow the employment of each code transformation. The functionality agreements between the original and transformed kernels were verified. Then the performance, power, and energy results of the original and transformed kernels were compared to analyze the impacts of the transformations. Several code transformations were investigated. Due to space constraints, we focus on the results for array declaration sorting, loop peeling, and procedure integration.

4.5 CODE OPTIMIZATION IMPACT ON POWER CONSUMPTION

In this section, we explore and analyze the experimental results of evaluating the impacts of various optimization techniques on power consumption.

4.5.1 IMPACT OF GLOBAL COMPILER OPTIMIZATION

This section discusses the different compiler optimization levels offered by the CCS3.1 along with the individual optimization tasks performed at each level. The targeted C/C++ compiler features many levels of optimization tuned to achieve speed by invoking the -o0, -o1, -o2, and -o3 options for global speed optimization [28].

We start by evaluating the effects of the four global compiler optimization levels (-o0, -o1, -o2, -and o3) on the energy and power consumption of the targeted processor. Moreover, we analyzed the effects on other performance measures such as memory references and cache miss rates. Figure 4.1 presents the measured power consumptions and averages at the four global performance optimization levels for different signal and image processing benchmarks. It is obvious from the figure that these optimization options generally increase power consumption.

The most aggressive optimizations they may lead to are minimum execution times but do not necessarily achieve the best code for minimal power consumption.

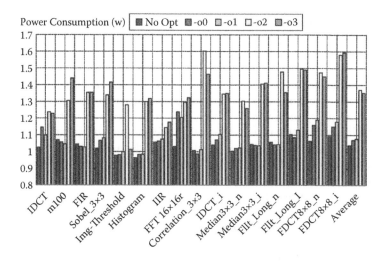

FIGURE 4.1   Power consumption of C6416T while running different benchmarks.

TABLE 4.2

Effects of Employing Global Optimization Options on Power, Energy, and Execution Time

Optimization	Power (%)	Execution Time (%)	Energy (%)
-o0	3.11	−27.35	−27.99
-o1	3.841	−22.32	−23.58
-o2	32.16	−91.17	−89.06
-o3	30.36	−96.20	−94.85

The highest optimization level (-o3) increased power consumption on average by 30.3% compared to the no-optimization option [29]. This percentage reached 45% for two individual benchmarks (FDCT8×8_i and correlation_3×3). On average, invoking -o2 or -o3 led to greater power consumption than invoking -o0 or -o1.

The software pipelining loop feature enabled with -o2 and -o3 allowed better instruction parallelization. As we will explain later, this feature had a significant impact on power consumption.

Although the results in Figure 4.1 demonstrate that invoking global optimization levels increases power consumption on average, the energy significantly decreased. Table 4.2 demonstrates that there is a strong correlation between execution time and energy consumption.

The most aggressive speed optimization level (-o3) reduced execution time on average by 96.2% compared to the no-optimization option. While −o3 reduces the energy on average by 94.8%, it is obvious that invoking -o2 and -o3 saves more energy than invoking -o0 or -o1. This can be explained by the enabling of the Software Pipelined Loop (SPLoop) at -o2 and -o3 that leads to far greater reduction in execution time and hence in energy use. The effect of SPLoop will be covered later in this chapter.

Based on the results shown in Table 4.2 we can distinguish two groups of optimization levels (-o0, -o1) improved by register allocation, and (-o2, -o3) discriminated by the use of SPLoop). In the (-o0, -o1) group, data were fetched simultaneously with the instruction execution, saving CPU cycles, thus consuming less energy. However, in the case of no optimization, data were pre-fetched from memory before execution of the instruction that required these data. If a program utilizes many variables, the power consumption increases while the energy decreases. Conversely, if a program utilizes few variables it retains an energy decrease through unchanged power consumption because the registers are loaded less frequently and reused more often.

The (-o2,- o3) group is characterized mainly by software pipelined loops. SPLoops save a fixed number of cycles per loop iteration, mostly by avoiding pipeline stalls. Since stalls consume less power than normal instruction executions, shortening the programs this way actually increases power consumption. However, the magnitude of the increase depends on how long the loop kernel is and the instructions it carries. To a lesser extent, the elimination of global common subexpressions further reduces execution time and power consumption.

To analyze the previous results, we found it worthwhile to study the effects of compiler optimizations on four important execution characteristics: L1 data cache misses, memory references, instructions per cycle (IPCs), and CPU stall cycles. Table 4.3 illustrates the effects of different optimization levels on the four characteristics. All the percentage changes in Table 4.3 are obtained with respect to the case of no optimization.

The L1D cache misses decreased on average by almost 69% when -o3 was invoked. The L1D cache misses required the access of the L2D cache/SRAM which in turn required additional power consumption. The CPU stall cycles decreased by 78% when -o3 was invoked. Several reasons such as cache misses, resource conflicts, and memory bank conflicts can cause a CPU to stall. Although one data cache miss caused at least six CPU stall cycles, two features of the C6416T CPU are expected to decrease cache miss penalties.

First, the L1D cache of the C6416T DSP pipelines the L1D cache read misses. A single L1D read miss takes six cycles when serviced from L2/SRAM and eight cycles when serviced from the L2 cache. Pipelining of cache misses can hide many miss

TABLE 4.3

Effects of Employing Global Optimization Options on Execution Characteristics

Optimization	L1D Cache Misses (%)	CPU Stall Cycles (%)	IPCs (%)	Memory References (%)
-o0	0.0	2.09	59.28	−81.89
-o1	−1.12	−1.29	59.14	−81.91
-o2	−1.12	−45.05	256.5	−82.69
-o3	−69.27	−78.75	269.95	−94.12

penalties (CPU stall cycles) by overlapping the processing of several cache misses. The miss overhead can be expressed as $[4 + (2 \times M)]$ when serviced from L2/SRAM or as $[6 + (2 \times M)]$ when serviced from L2 (M is the number of cache misses) [30]. The pipelining of cache misses significantly reduced execution time and energy but exerted no effect on power consumption.

Second, the write cache miss feature does not directly stall the CPU because of the use of L1D write buffer [30]. This feature also affects execution time but has no effect on power consumption especially when the write buffer is not full.

Table 4.3 indicates that IPCs increased by about 269% when -o3 was invoked compared to the case when all optimization options are disabled. This surely decreases execution time and energy because more overlapping in the execution of the instructions per cycle is achieved. However, the resulting higher parallelization increases the power consumption.

Although Table 4.3 indicates that the memory references decreased by 94% when -o3 was invoked (and expected to save power), we found that the power use increased. This emphasizes our results [31,32] demonstrating that the instruction management unit (IMU) responsible for fetching and dispatching instructions dominates the memory referencing contribution.

4.5.2 Effects of Specific Architectural Features

In this section, we explore the impacts of two special architectural features of the targeted DSP: the SPLoop and SIMD on power consumption.

4.5.2.1 Impact of Software Pipelined Loop

The SPLoop, also called the hardware zero overhead loop (ZOL) is a type of instruction scheduling that exploits instruction level parallelism (ILP) across loop iterations. SPLoop is a specific architectural optimization feature of the C64x+ CPU; the C64x CPU does not support SPLoop. This feature allows the CPU to store a single iteration of a loop in a specialized buffer containing hardware that will selectively overlay copies of the single iteration in a software pipeline to construct an optimized execution of the loop [33].

Modulo scheduling is a form of SPLoop that initiates loop iterations at a constant rate called the iteration interval (ii). To construct a modulo scheduled loop, a single loop iteration is divided into a sequence of stages, each with length ii. In the steady state of the execution of the SPLoop, each stage executes in parallel. The instruction schedule for a modulo scheduled loop has three components: a prolog, a kernel, and an epilog. The kernel is the instruction schedule that executes the pipeline steady state. The prolog and epilog are instruction schedules that set up and drain the execution of the loop kernel [33].

This section evaluates the impact of software pipelining on power and energy consumption. The SPLoop feature is implicitly enabled with global optimization options -o2 and -o3. However, we overrode this by invoking the -mu option in conjunction with -o2 or -o3 to disable only the SPLoop. To distinguish whether software pipelining is enabled or disabled, we used -o2-mu and -o3-mu to indicate disabling of the SPLoop.

TABLE 4.4

Average Power, Execution Time, and Energy for Investigated Benchmarks

Benchmark	Power (W)	Execution Time (ms)	Energy (mJ)
-o2	1.371	0.168	0.221
-o2-mu	1.109	0.703	0.766
-o3	1.352	0.072	0.104
-o3-mu	1.117	0.16	0.204

Table 4.4 summarizes the average power, execution time, and energy for different signal and image processing benchmarks when -o2, -o2-mu, -o3, and -o3-mu were invoked. The table clearly illustrates the strong impact of the SPLoop on execution time. When the -o2-mu was invoked, the execution time increased by 317.3% relative to -o2. Execution time increased by 120.8% when -o3-mu was invoked relative to -o3. Clearly the impact of disabling the SPLoop on execution cycles was greater with -o2 than with -o3. The -o3 option includes all the individual optimizations in -o2 along with more performance-oriented optimizations intended to reduce execution cycles through reductions in memory references.

Despite the increase in the execution cycles when SPLoop was disabled, the power consumption decreased on average by 19.1 and 17.4%, respectively, when -o2-mu and -o3-mu were invoked. Table 4.4 demonstrates that the energy increases when SPLoop is disabled and the increase relates directly to the increase in execution time.

Power consumption increased on average by 7.67% when -o3-mu was invoked compared to the case of no optimization option. The SPLoop contributed 70.3% to the total power increase when -o3 was invoked. More attention should focus on the design of specialized hardware for software pipelining to achieve a compromise of performance and power trade-offs for the C6416T.

Figure 4.2 summarizes the effects of the SPLoop on power consumption. It is pretty clear that invoking -o3-mu results in a tradeoff between performance and power consumption.

4.5.2.2 Impact of SIMD

The C6000 compiler recognizes a number of intrinsic C-functions. Intrinsics allow a programmer to express the meanings of certain assembly statements that would otherwise be cumbersome or inexpressible in C/C++. Most intrinsic functions utilize the SIMD capabilities of the C6416T. Intrinsics are used like functions. A programmer can use C/C++ variables with these intrinsics, just as he or she would with any normal function. The intrinsics are specified with a leading underscore and are accessed by calling them as function.

For example:

```
int X1, X2, Y;
Y = _sadd(X1, X2)
```

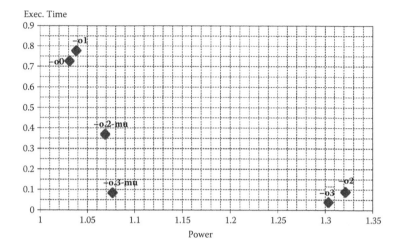

FIGURE 4.2   Execution time versus power consumption at various optimization levels.

Original Code	Code with Intrinsics

```
1 #include "idct_8x8_c.h"
2 #pragma CODE_SECTION(idct_8x8_cn,  ".text:ansi");
3 void idct_8x8_cn(short *idct_data, unsigned num_idcts)
4 {
5     _nassert((int) idct % 8 == 0);
6     _nassert(num_idcts >= 1);
7     for (i = 0; i < num_idcts; i++)
8     {
9         for (j = 0; j < 8; j++)
10        {
11            F0 = idct[i][0][j];
12            F1 = idct[i][1][j];
13            F2 = idct[i][2][j];
14            F3 = idct[i][3][j];
15            F4 = idct[i][4][j];
16            F5 = idct[i][5][j];
17            F6 = idct[i][6][j];
18            F7 = idct[i][7][j];
19
20            P0 = F0;            P1 = F4;
21            R1 = F2;            R0 = F6;
22
23            Q1 = (F1*C7 - F7*C1 + 0x8000) >> 16;
24            Q0 = (F5*C3 - F3*C5 + 0x8000) >> 16;
25            S0 = (F5*C5 + F3*C3 + 0x8000) >> 16;
26            S1 = (F1*C1 + F7*C7 + 0x8000) >> 16;
27
28            p0 = ((int)P0 + (int)P1 + 1 ) >> 1;
29            p1 = ((int)P0 - (int)P1     ) >> 1;
30            r1 = (R1*C6 - R0*C2 + 0x8000) >> 16;
31            r0 = (R1*C2 + R0*C6 + 0x8000) >> 16;
```

```
1 #include "idct_8x8_i.h"
2 #pragma CODE_SECTION(idct_8x8_c,  ".text:intrinsic");
3 void idct_8x8_c(short *idct_data, unsigned num_idcts)
4 {
5     _nassert((unsigned)idct_data % 8 == 0);
6     #pragma MUST_ITERATE(4,,4);
7     for (i = jC = j1 = 0; i < num_idcts * 4; i++)
8     {
9         F00 = _amem4(&_ptr[ 0 + 2*j0]);
10        F11 = _amem4(&_ptr[ 8 + 2*j0]);
11        F22 = _amem4(&_ptr[16 + 2*j0]);
12        F33 = _amem4(&_ptr[24 + 2*j0]);
13        F44 = _amem4(&_ptr[32 + 2*j0]);
14        F55 = _amem4(&_ptr[40 + 2*j0]);
15        F66 = _amem4(&_ptr[48 + 2*j0]);
16        F77 = _amem4(&_ptr[56 + 2*j0]);
17
18        if (++j0 == 4; ( j0 = 0; i_ptr += 64; )
19
20        F17 = _pack2(F11, F77);
21        F53 = _pack2(F55, F33);
22        F26 = _pack2(F22, F66);
23        F04 = _pack2(F00, F44);
24
25        Q1 = _dotpnsu2(F17, C71);
26        Q0 = _dotpnsu2(F53, C35);
27        S0 = _dotprsu2(F53, C53);
28        S1 = _dotprsu2(F17, C17);
29
30        S0Q0 = _pack2(S0, Q0);
31        S1Q1 = _pack2(S1, Q1);
```

FIGURE 4.3   Example of IDCT kernel with and without SIMD utilization.

For a complete list of the C6000 and specific C64x+ intrinsic functions readers are encouraged to review Reference [28].

To assess the effects of utilizing SIMD instructions on energy and power consumption, we prepared two versions of the inverse discrete cosine transform (IDCT) algorithm as a case study. The first version was implemented without SIMD instructions; the second was implemented with all possible SIMD instructions as shown in Figure 4.3. The functionalities of both versions were tested and verified to yield the same results.

We studied the effect of the employing SIMD instructions isolated from the effects of the SPLoop feature by compiling two versions with -o2-mu and -o3-mu

(-mu disables the SPLoop feature). It is worth mentioning that invoking -o0 or -o1 will not enable the SPLoop.

Table 4.5 demonstrates that the SIMD version of the IDCT compiled and optimized with -o3-mu achieved a 3.96% power saving along with 25.4 and 28.35% reductions in execution time and energy, respectively. The achieved power saving is mainly the result of reduction of the IPC by 20.86%. The enhancement of execution time was derived by a significant reduction (more than 62%) on memory references.

To summarize the effects of utilizing SIMD on power and energy consumption, we conducted two more case studies utilizing a discrete cosine transform (DCT) and a median filter with a 3 × 3 window in the same manner as the IDCT investigation. Figures 4.4 and 4.5 represent comparisons of power consumption and energy with and without SIMD at various performance optimization levels.

In general, employing SIMD significantly enhanced performance and energy saving. The SPLoop was the main basis for the significant improvement in performance when -o2 or -o3 was invoked [35]. By disabling the SPLoop feature, -o2-mu or -o3-mu, the utilization of SIMD produced a comparable performance enhancement

TABLE 4.5

Effects of Employing SIMD While Invoking -o3-mu Optimization

	Original	With Intrinsics	%
Execution cycles	3319	2476	−25.4
Power (W)	1.091	1.048	−3.96
Energy (mJ)	0.00362	0.00259	−28.35
IPCs	2.416	1.913	−20.86
CPU stall cycles	96	0	−100
Memory references	1536	576	−62.5

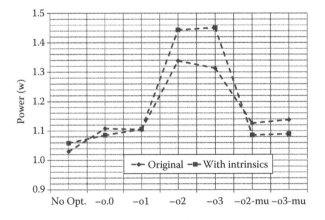

FIGURE 4.4 Power consumption with and without SIMD utilization versus various optimization options.

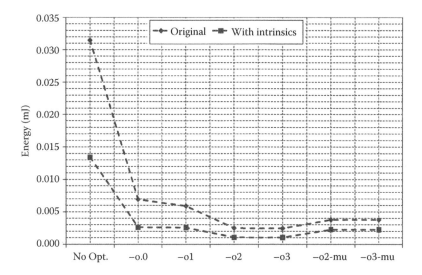

FIGURE 4.5   Energy with and without SIMD utilization versus various optimization options.

with -o2 or -o3 along with the great advantage of average power savings of 18.83 and 17%, respectively.

It is clear that rewriting the algorithm to yield maximum utilization of SIMD instructions while invoking the −o3 optimization options was the best choice from power consumption and performance perspectives. This choice may be considered a tradeoff between power consumption on one side and execution time and energy use on the other.

4.5.3 INFLUENCE OF SOURCE-TO-SOURCE CODE TRANSFORMATIONS

Code reshaping techniques, also known as source-to-source code transformations, consist of loop and data flow transformations. They are essential to modern optimizing and parallelizing compilers. They enhance temporal and spatial locality for cache performance and expose an algorithm's inherent parallelism to loop nests [21].

In this section, we assess the impacts of applying source to source code transformations from power and energy perspectives. The source code transformations presented in this chapter are classified into loop, data, and procedural transformation groups.

To evaluate the effectiveness of the applied transformations, we compiled the original and transformed versions of each program on the target architecture. We recorded the current drawn (consumed power) from the core CPU. With the aid of a CCS3.1 profiler, we recorded execution times and other execution characteristics such as memory references, L1D cache misses, and others. To obtain reliable and precise information, we repeated the measuring procedure several times for each transformation [32].

First we assessed the effects of loop peeling transformations. This type of transformation, also called loop splitting, attempts to eliminate or reduce loop dependencies introduced in a few first or last iterations by splitting them from the loop and performing them outside the loop to allow better instruction parallelization. Moreover, the transformation can be used to match the iteration control of adjacent loops, allowing two loops to be fused together.

Figure 4.6 shows an example of loop peeling transformation. In the original code of this example, the first iteration makes use only of the variable $p = 10$; for all other iterations, the variable is $p = i - 1$. Therefore, the first iteration in the transformed code is moved outside the loop and the loop iteration control is modified.

Table 4.6 shows the impact of applying the loop peeling transformation on power, energy, and execution time. Because of splitting the first iteration from the loop's body and performing it outside the loop, the memory references decreased by 37.78% while maintaining the same number of L1D cache misses. The instruction parallelization expressed as IPCs, improved by 4.6%. The execution time and power consumption were enhanced by 11.5 and 2.78%, respectively, leading to an energy saving of 13.97%.

Second, we present array declaration sorting transformation as an example of data-oriented transformations. Figure 4.7 shows an example in which the array access frequency ordering is C[], B[] and A[]. The A[], B[], and C[], declaration order in the original code is restructured placing C[] in the first position, B[] in the second, and A[] at the end. This declaration reordering is employed to assure that frequently accessed arrays are placed on top of the stack. In this way, the frequently used memory locations are accessed by exploiting direct access mode.

Original Code	Transformed Code
| ```
int p = 10;
for (i=0; i<N; ++i)
{
    y[i] = x[i] + x[p];
    p = i;
}
``` | ```
y[0] = x[0] + x[10];
for (i=1; i<N; ++i)
{
    y[i] = x[i] + x[i-1];
}
``` |

FIGURE 4.6 Loop peeling transformation.

TABLE 4.6

Effects of Loop Peeling Transformation on Energy and Power Consumption

	Original	Transformed	%
Execution cycles	2808	2485	−11.5
Power (W)	1.034	1.006	−2.78
Energy (mJ)	0.0029	0.0025	−13.97
IPCs	0.919	0.962	4.6
Memory references	802	499	−37.78

Original Code	Transformed Code
```int A[DIM], B[DIM], C[DIM], i;```	```int C[DIM], B[DIM], A[DIM], i;```

```
int A[DIM], B[DIM], C[DIM], i;
    for (i = 5; i < 3500; i+=5)
        C[i] = val;
    for (i = 5; i < 2000; i+=10)
        B[i] = val;
    for (i = 5; i < 1000; i+=10)
        A[i] = val;
}
```

```
int C[DIM], B[DIM], A[DIM], i;
    for (i = 5; i < 3500; i+=5)
        C[i] = val;
    for (i = 5; i < 2000; i+=10)
        B[i] = val;
    for (i = 5; i < 1000; i+=10)
        A[i] = val;
```

FIGURE 4.7   Array declaration sorting transformation.

The array declaration sorting reduces execution time by 1.95% and consequently saves energy by 2.19%. Power consumption is little affected so this transformation is not power hungry.

Third, we studied the impact of procedure integration, also known as procedure inlining. Procedure integration replaces calls to procedures with copies of their bodies [36]. It can be a very useful optimization because it changes calls from opaque objects that may exert unknown effects on aliased variables and parameters to local code that exposes its effects and may be optimized as part of the calling procedure [37].

Although procedure integration removes the costs of the procedure call and return instructions, the savings are often small. The major savings arise from the additional optimizations that become possible on the integrated procedure body. For example, a constant passed as an argument can often be propagated to all instances of the matching parameter. Moreover, the opportunity to optimize integrated procedure bodies can be especially valuable if it enables loop transformations that were originally inhibited by embedding procedure calls in loops or turning a loop that calls a procedure (whose body is itself a loop) into a nested loop [37].

Ordinarily, when a function is invoked, the control is transferred to its definition by a branch or call instruction. With procedure integration, the control flows directly to the function code without a branch or call instruction. Moreover, the stack frames for the caller and called are allocated together. Procedure integration may make the generated code slower as well, for example, by decreasing the locality of the reference. Figure 4.8 shows an example of procedure integration. The *pred(int)* function is integrated into the *f(int)* function.

Table 4.7 shows the impacts of applying procedure integration transformations on power, energy, and execution time. As noted earlier, procedure integration eliminates call overhead and consequently reduces the memory references in the proposed example by 12.44%. Moreover, procedure integration reduces the executed instructions by 41.11% and IPCs by 12.59%. Thus, power consumption and execution time are reduced by 3.93 and 32.63%, respectively.

Finally, Figure 4.9 summarizes the results of applying different code transformations to power, execution time, and energy. The original code represents 100%. Any deviation above or under 100% is related to the applied code transformation.

Original Code	Transformed Code
| ```
main()
{
int i,res[DIM] , val = 7;
    for( i = 0; i < N; i++ )
    {
        res[i] = f(val);
        val += 5;
    }
}

int f(int y)
{
    return pred(y) + pred(0)
            + pred(y+1);
}

int pred(int x)
{
    if (x == 0)
        return 0;
    else
        return x-1;
}
``` | ```
main()
{
int i,res[DIM] , val = 7;
    for( i = 0; i < N; i++ )
    {
        res[i] = f(val);
        val += 5;
    }
}

int f(int y)
{
    int temp = 0;
    if (y == 0) temp += 0;
    else temp += y - 1;
    if (0 == 0) temp += 0;
    else temp += 0 - 1;
    if (y+1 == 0) temp += 0;
    else temp += (y+1) - 1;
    return temp;
}
``` |

FIGURE 4.8 Procedure integration transformation.

TABLE 4.7

Effects of Procedure Integration Transformations on Energy and Power Consumption

	Original	Transformed	%
Execution cycles	3218	2168	−32.63
Power (W)	1.039	0.998	−3.93
Energy (mJ)	0.0033	0.0022	−35.27
IPCs	0.983	0.859	−12.59
Memory references	804	704	−12.44
Executed Instructions	3162	1862	−41.11

The results show that several code transformations such as loop peeling, loop fusion, and procedure integration produce good impacts on power consumption, energy, and performance. Other transformations such as loop permutation and loop tiling improve power consumption through performance. The results also show that some transformations have no impacts on power consumption but improve performance and energy. These loop reversal, loop strength reduction, and array declaration sorting transformations are not power hungry. The last group of code transformations (loop unswitching, loop normalization, fusion, and scalarization of array elements) improve the performance through power consumption.

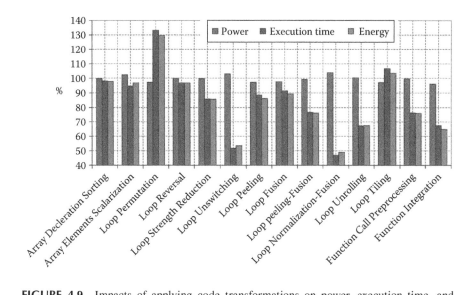

FIGURE 4.9 Impacts of applying code transformations on power, execution time, and energy.

4.6 CONCLUSIONS

Reducing energy and power dissipation of an embedded system have become optimization goals in their own right. They are no longer considered side effects of traditional performance optimizations that attempt to reduce program execution time. In this chapter, we investigated the impacts of a set of high level language optimization techniques on the power consumption levels of embedded system architectures.

As a specific example, we considered the powerful Texas Instruments C6416T DSP processor. We presented a qualitative study that evaluated global compiler-based optimization levels (-o0, -o1, -o2, and -o3), certain architectural features, and several source code transformations from energy and power consumption perspectives.

First, we precisely measured the impacts of applying global execution time optimization levels on power consumption. The results indicated that the most aggressive execution time optimization level -o3 increased consumption by 30.3%. Meanwhile, energy decreased by 94.8%. This was mainly due to aggressive utilization of instruction execution overlapping that reduced execution time and energy but not power consumption.

Second, we assessed the effect of the C64x+ architectural feature known as SPLoop on energy and power consumption. The results show that the software loop pipelining feature contributed on average 69.7% to the total power consumption increase if -o3 was invoked.

We also investigated the influence of the powerful capability of the TMS320C6416T to execute SIMD instructions. We investigated the effect of SIMD while enabling and disabling SPLoop. By disabling the SPLoop, -o2-mu, or -o3-mu, SIMD instruction use generated comparable performance enhancement with -o2 or -o3 along with great advantages of 18.83 and 17% average power savings, respectively.

Finally, we examined the impacts of applying source code transformations on energy and power consumption. Three main categories of code transformations (data, loop, and procedural oriented) were investigated. Several code transformations (loop peeling, loop fusion, and procedure integration) demonstrated good impacts on power consumption, energy, and performance.

Loop permutation and loop tiling improved power consumption on account of performance. The results also indicated that some transformations produced no impacts on power consumption but they improved performance and decreased energy consumption. Loop reversal, loop strength reduction, and array declaration sorting transformations are not power hungry.

REFERENCES

1. D. Ortiz and N. Santiago. 2007. High level optimization for low power consumption on microprocessor-based systems. In *Proceedings of 50th IEEE Midwest Symposium on Circuits and Systems*, Montreal Canada, pp. 1265–1268.
2. M. Valluri and L. John. 2001. Is compiling for performance == compiling for power? In *Proceedings of 5th Workshop on Interaction between Compilers and Computer Architectures*, Monterrey, Mexico.
3. Standard Performance Evaluation Corporation (SPEC). n.d. *SPEC Benchmark Suite.* http://www.spec.org
4. R.A. Ravindran, P.D. Nagarkar, G.S. Dasika et al. 2005. Compiler managed dynamic instruction placement in a low-power code cache. In *Proceedings of International Symposium on Code Generation and Optimization*. San Jose, California, pp. 179–190.
5. L.N. Chakrapani et al. 2001. The emerging power crisis in embedded processors: what can a poor compiler do? In *Proceedings of ACM International Conference on Compilers, Architecture, and Synthesis for Embedded Systems* Atlanta, Georgia, USA, pp. 1–6.
6. J.S. Seng and D.M. Tullsen. 2003. The effect of compiler optimizations on Pentium 4 power consumption. In *Proceedings of 7th IEEE CS Workshop on Interaction between Compilers and Computer Architectures*. Anaheim, CA, USA, pp. 51–56.
7. J. Zambreno, M.T. Kandemir and A.N. Choudhary. 2002. Enhancing compiler techniques for memory energy optimizations. In *Proceedings of 2nd International Conference on Embedded Software*. Grenoble, France, Springer-Verlag, pp. 364–381.
8. M. Casas-Sanchez, J. Rizo-Morente, C. Bleakley et al. 2007. Effect of compiler optimizations on DSP processor power and energy consumption. In *Proceedings of Conference on Design of Circuits and Integrated Systems*, Seville, Spain.
9. G. Esakkimuthu, N. Vijaykrishnan, M. Kandemir et al. 2000. Memory system energy: influence of hardware–software optimizations. In *Proceedings of ACM International Symposium on Low Power Electronics and Design*. New York, pp. 244–246.
10. C. Lee, J. K. Lee, T. Hwang et al. 2000. Compiler optimization on instruction scheduling for low power. In *Proceedings of 13th IEEE International Symposium on System Synthesis*, Madrid, Spain, pp. 55–60.
11. M. Kandemir, N. Vijaykrishnan, and M.J. Irwin. 2002. Compiler optimizations for low power systems. In R. Graybill and R. Melhem, Eds., *Power-Aware Computing*, pp. 191–210.
12. W. Ye, N. Vijaykrishnan, M. Kandemir et al. 2000. The design and use of SimplePower: a cycle-accurate energy estimation tool. In *Proceedings of 37th ACM/IEEE Conference on Design Automation*. Los Angeles, pp. 340–345.
13. H. Mehta, R.M. Owens, M.J. Irwin et al. 1997. Techniques for low energy software. In *Proceedings of ACM/IEEE International Symposium on Low Power Electronics and Design*. Monterey, California, USA, pp. 72–75.

14. R. Leupers. 2000. Code generation for embedded processors. In *Proceedings of 13th ACM/IEEE International Symposium on System Synthesis*. Madrid, Spain, pp. 173–178.
15. J. Oliver, O. Mocanu, and C. Ferrer. 2003. Energy awareness through software optimization as a performance estimate case study of the MC68HC908GP32 microcontroller. In *Proceedings of 4th IEEE International Workshop on Microprocessor Testing and Verification*. Austin, Texas, USA, pp. 111–116.
16. M.T.C. Lee, M. Fujita, V. Tiwari et al. 1997. Power analysis and minimization techniques for embedded DSP software. *IEEE Transactions on VLSI Systems*, 5, 123–135.
17. D. Ortiz and N. Santiago. 2008. Impact of source code optimizations on power consumption of embedded systems, In Proceedings of the 6th IEEE International Joint Conference on Circuits and Systems and TAISA NEWCAS-TAISA. Montreal, Canada, pp. 133–136.
18. Z.N. Azeemi and M. Rupp. 2005. Energy-aware source-to-source transformations for a VLIW DSP processor. In *Proceedings of the 17th International Conference on Microelectronics*. Islamabad, Pakistan, pp. 133–138.
19. C. Brandolese, W. Fornaciari, F. Salice et al. 2002. The impact of source code transformations on software power and energy consumption. *Journal of Circuits, Systems, and Computers*, 11, 477–502.
20. D. Burger and T.M. Austin. 1997. The SimpleScalar Tool Set, Version 2.0. *Computer Architecture News*, 25, 13–25.
21. F. Catthoor, K. Danckaert, S. Wuytack et al. 2001. Code transformations for data transfer and storage exploration preprocessing in multimedia processors. *IEEE Design and Test of Computers*, 18, 70–82.
22. C. Kulkarni, F. Catthoory, and H. De Man. 1998. Code transformations for low power caching in embedded multimedia processors. In *Proceedings of 12th. IEEE CS International Symposium on Parallel Processing*. Orlando, Florida, pp. 292–297.
23. K.S. McKinley, S. Carr, and C.W. Tseng. 1996. Improving data locality with loop transformations. *ACM Transactions on Programming Languages and Systems*, 18, 424–453.
24. H. Yang, G.R. Gao, A. Marquez et al. 2001. Power and energy impact by loop transformations. In *Proceedings of Workshop on Compilers and Operating Systems for Low Power: Parallel Architecture and Compilation Techniques*. Barcelona, Spain, pp. 12.1–12.8.
25. Texas Instruments Inc. 2003. *TMS320C6416T Fixed Point Digital Signal Processor Datasheet*. www.ti.com
26. Spectrum Digital. 2004. *TMS320C6416T,DSK Technical Reference*. http://c6000. spectrumdigital.com/dsk6416/
27. Agilent Technologies Inc. 2007. *Agilent 34410A Digital Multimeter Datasheet*. http://www.home.agilent.com/agilent/product.jspx?pn = 34410A
28. Texas Instruments Inc. 2004. *TMS320C6416T Fixed Point Digital Signal Processor Optimizing Compiler User Guide*. www.ti.com
29. M.E.A. Ibrahim, M. Rupp, and S.E.D. Habib. 2009. Compiler-based optimizations impact on embedded software power consumption. In *Proceedings of IEEE Joint Conference Circuits and Systems and TAISA*, (NEWCAS-TAISA'09), Toulouse, France, pp. 247–250.
30. Texas Instruments Inc. 2006. *TMS320C6416T DSP Two-Level Internal Memory Reference Guide*. www.ti.com
31. M.E.A. Ibrahim, M. Rupp, and H.A.H. Fahmy. 2008. Power estimation methodology for VLIW digital signal processor. In *Proceedings of IEEE Conference on Signals, Systems and Computers*. Asilomar, CA, USA, pp. 1840–1844.
32. M.E.A. Ibrahim, M. Rupp, and H. A. Fahmy. 2011. A precise high-level power consumption model for embedded systems software. *EURASIP Journal on Embedded Systems*, vol. 2011.

33. Texas Instruments Inc. 2007. *TMS320C64x/C64x+ DSP CPU and Instruction Set Reference Guide.* www.ti.com

34. Texas Instruments Inc. 2009. *C6000 Host Intrinsics.* www.tiexpressdsp.com

35. M.E.A. Ibrahim, M. Rupp, and S.E.D. Habib. 2009. Performance and power consumption trade-offs for a VLIW DSP. In *Proceedings of IEEE International Symposium on Signals, Circuits and Systems.* Iasi, Romania, pp. 197–200.

36. D.F. Bacon, S.L. Graham, and O.J. Sharp. 1994. Compiler transformations for high-performance computing. *ACM Computing Surveys*, 26, 421–461.

37. S.S. Muchnick. 1997. *Advanced Compiler Design and Implementation.* San Francisco: Morgan Kaufmann.

5 Flexible and Scalable Power Management Framework for RTOS-Based Embedded Systems

Muhamed Fauzi, Thambipillai Srikanthan, and Siew-Kei Lam

CONTENTS

5.1 INTRODUCTION

Rising energy costs and consumer demands for longer use have become primary concerns for portable embedded devices. In addition, battery technology progresses slowly, improving by only a few percent each year. Portable device sizes continue to shrink and this restricts the amount of space for battery storage. The energy concerns and increasing market pressures for embedded systems therefore necessitate novel strategies for rapid incorporation of power saving features and integration of new power management policies.

As embedded systems became more complicated, the complexity of the underlying hardware has been affected by the inclusion of real-time operating system (RTOSs). Previous work has demonstrated that system-wide energy consumption [2] can be managed effectively by RTOS. Although a large amount of work in OS-based power management [2] has been reported, little has been done to address its applicability in embedded systems. Recent efforts focusing on OS-directed power management lack application specificity and overlook real-time considerations of embedded systems. In addition, existing RTOS have limited power and proprietary management features.

This research work is motivated by the need for a power management framework that provides well defined interfaces between RTOS and devices to facilitate rapid incorporation of power strategies. These strategies are capable of adapting to the workload characteristics of an application while meeting the constraints of embedded systems. We proposed an ACPI-like power management framework for embedded systems. The proposed framework features a common API for different RTOSs and target platforms. This enables policy developers to incorporate target-independent power management policies for application and RTOS. This overcomes the portability restrictions of current power management policies.

The proposed framework enables the rapid development of power management algorithms and provides a means for the RTOS to be well informed about current power status of a device. This information can be exploited for making optimal power management decisions to coordinate energy resources among multiple tasks and adapt system operations according to task-specific requirements. The framework facilitates the development of future power policy algorithms tailored for different situations and provides a means to decouple development of applications and power management policies.

5.2 RELATED WORK

The QNX Neutrino power management architecture was designed to allow developers to build fully customized power management solutions specifically for their target hardware. The architecture makes no assumptions about specific power management standards. Instead it allows developers to have full control over the power resources of their systems [9]. QNX Neutrino is a hard real-time system that incorporates a power management framework into the RTOS and requires the application to be migrated to the RTOS to take full advantage of the power features.

VxWorks power management strategy is tightly coupled with the microcontroller. Power management [10] capabilities are provided via libraries used to access the microcontroller's power management features outside the control of the RTOS.

Power management in Android is built around the Linux power management module. Android uses wake locks and time-out mechanisms to switch system power states. Wake locks are used by applications and services to request microcontroller resources. If no wake locks are active, Android will shut down the microcontroller [3].

The advanced configuration and power interface (ACPI) is a popular power management framework used in desktop computers. ACPI evolved from advanced power management (APM) to fill the need for better power management. OS manages power unlike the BIOS in APM that improves system stability [1]. ACPI is a specification that standardizes the description of power management capabilities to facilitate the OS to manage power for the system. Power states are defined at various granularities of consumption for individual devices and entire systems. With a whole gamut of power states to choose, the onus is on the power management policy to determine the optimal power state for a system [4,6].

5.2.1 Advantages of Proposed Framework

Similar to our framework, QNX's version can manage all the devices on a platform. It utilizes a device driver that must be modified to include specific power functions. Our framework reduces the engineering effort needed to update the RTOS on power methods and data of the platform by incorporating ASL script to describe the target platform power features.

VxWorks also utilizes key concepts of ACPI specification for its power management framework. Similar to Android, the VxWorks power manager focuses only on the CPU to reduce power usage and like the QNX Neutrino, the VxWorks is embedded into the RTOS. Our framework has been shown to adapt by offering alternative designs to complement specific RTOS.

Android supports its own power management on top of the standard Linux power management and is limited to using wake locks to request CPU resources. Our framework handles all devices on the platform including the CPU as power-consumers. In addition, our framework can incorporate advanced power policy algorithms tailored to an application to manage the power of a platform. The proposed power management framework also features the following advantages:

Hardware abstraction—The hardware is described independently from the RTOS native language (in the form of ASL tables). Any hardware change requires only updating the ASL table (file). The framework can work without dependency on specific hardware configuration of the system.

Policy abstraction—Power policy defines the rules by which the system hardware devices transition among various power states. The framework only provides the infrastructure for the policy manager to implement its decisions.

Standardized middleware for power management—The framework is designed to be platform independent and thus can be deployed onto most platforms with minimal changes. This feature is also seen in ACPI which provides a similar capability for computer systems.

Support for dynamic power management—The framework is capable of facilitating dynamic power management for embedded systems. It does not assume any static set of policies or timed power transition models. The framework is also designed to take runtime requests from the policy and act accordingly.

Scalability—Changes to the hardware only require details composed of the data and methods necessary to control the platform power features in a script-like language. Another facet of scalability is its ability to change the power policy easily to fit the application.

5.3 PROPOSED RTOS POWER MANAGEMENT FRAMEWORK

The proposed framework (Figure 5.1) exerts sole control of all power-related functions of a system. It offers a flexible infrastructure that allows policy algorithm to incorporate intelligent power management into the system. Standard interfaces are defined and provided for interactions between system components such as the target platform, RTOS, and policy algorithm. It is designed to be RTOS-agnostic and portable to any target platform.

The framework implements the RTOS-ACPI subsystem that extends the functionality of the RTOS and enables the RTOS to manage power at a holistic level and reduce the dependence on an application to manage power. The *power policy algorithm* makes intelligent decisions based on total system activity and then uses the subsystem to implement its decisions and manage the devices. However, these decisions are subject to the *basic policy manager* that prevents compromise of the real-time constraints of the system.

5.3.1 RTOS-ACPI SUBSYSTEM

The subsystem takes in an *ASL table (file)* that describes the platform's power features and power control methods. The *ASL translator* translates the information to a *Namespace* structure in the RTOS native format (Figure 5.2). The *Namespace* contains information like register addresses and device names, but does not contain

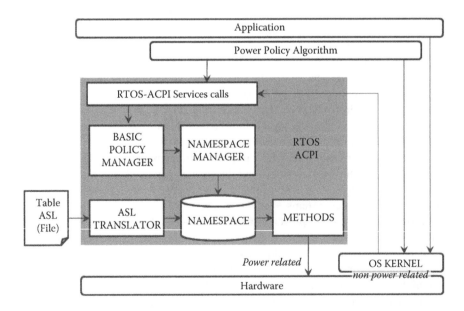

FIGURE 5.1   Proposed RTOS-ACPI framework.

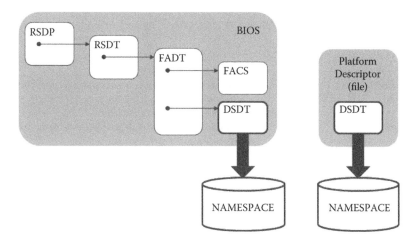

FIGURE 5.2   Difference in generating Namespace in ACPI and RTOS-ACP.

functions to control power; these are located in the *Methods*. The *Namespace manager* handles any requests to access the *Namespace*.

The subsystem interacts with the hardware directly because it has access to hardware-specific information (registers) in its Namespace and corresponding control methods found in Methods. Accessing an appropriate part of Namespace provides a gateway to the corresponding hardware devices. Most embedded platforms have on-chip devices that are easily accessible via register addresses. The following sections detail the subsystem composed of several modules. For the subsystem to work, the RTOS needs the device manager to notify and update the subsystem of any event affecting devices on the target platform.

5.3.2   RTOS-ACPI Service Calls

This module exposes the system calls needed by the application or *power policy algorithms* to utilize the RTOS-ACPI. It accepts these requests and validates them before giving access to *Namespace*.

The functions shown in Table 5.1 help the *power policy* obtain power-related information from the platform devices. Using this information, the *power policy* determines the next power state and utilizes the provided methods to achieve the desired power state. Some functions, such as *EnterSystemState*, *ChangeBusState*, and *EnterG0State* assist the policy to change the power state for the entire platform with a single system call.

Each call in the subsystem returns the E_OK code (shown in Table 5.2) if the control method is successful. However, if an error were to occur, the appropriate error code would be returned to the developer for corrective action.

5.3.3   ACPI Source Language (ASL) Tables

In ACPI, platform hardware developers provide the tables (DSDT, RSDT, etc.) that contain data and control methods pertaining to each device on the platform.

TABLE 5.1
RTOS-ACPI System Calls

System Call		Name and Function
ReadDeviceState	*description*	Reads device state from stored value, not current value
	Input	Device name
	return value	Device power state if successful
GetCurrentState	*description*	Obtains current state of device by reading device registers
	input	Device name
	return value	Device state if successful
GetIdleStart	*description*	Obtains idle start time
	input	Device name
	return value	Idle start time if successful
GetDeviceNames	*description*	Counts number of attached devices
	input	None
	return value	Number of attached devices if successful
DetStateTransTime	*description*	Time to transit device from one power state to another
	input	Device names for "from" and "to" states
	return value	Time in milliseconds if successful
CheckDeviceBusy	*description*	Determine whether device is being used
	input	Device name
	return value	Positive integer if successful (busy), 0 if not busy
ChangeDeviceState	*description*	Changes device state to specified power state
	input	Device names ("state to transit to")
	return value	E_OK indicates device transition successful
EnterSystemState	*description*	Transition of entire system to predefined custom state
	input	Typical state change from S0 to S4
	return value	E_OK indicates system transition successful
EnterG0State	*description*	Transition of entire system to wake state; all devices D0
	Input	None
	return value	E_OK if system transition successful
ChangeBusState	*description*	Transition all devices in bus to required state
	Input	Bus name ("state to transit to")
	return value	E_OK if all device transitions successful

TABLE 5.2
RTOS-ACPI Error Codes

Error Macro	Definition
E_OK	Success
E_PM_DISAPPROVE	RTOS-ACPI power manager disapproved
E_DEV_NOEXS	Device not present in platform
E_STATE_NOEXS	State not supported
E_DEV_NOATT	Device not attached
E_TT_UNKNOWN	Transition time unknown
E_ERR	General error
E_BUS_NOEXS	Bus not on system

The tables are stored in the BIOS and read by the operating system during start-up. Since BIOS is not commonly found in embedded platforms, our design provides the table information to the RTOS via a plaintext file. These tables are used to build a hierarchical data structure called *Namespace*.

Similar to ACPI, the tables are coded using ASL operators (shown in Table 5.3) to make them OS-independent. However, in this implementation, ASL code is not compiled to ACPI machine language (AML) code, but is translated to OS-specific data structures that are utilized at runtime. The ASL compiler and the AML interpreter have been discarded to significantly reduce complexity overhead.

The ASL is described in plaintext file using ASL syntax (shown in Figure 5.3)—not in assembly language nor OS native format. Reviewing the ACPI tables revealed that only the DSDT table contained vital power information and was incorporated into the plaintext file while omitting the other ACPI tables.

TABLE 5.3
RTOS-ACPI ASL Operators

ASL Operator	Description
Scope	Opens named scope
Device	Declares bus or device object
Processor	Declares processor
Name	Declares named object
Register	Declares register
External	Declares external object
Method	Declares control method
AND	Integer bitwise *and*
OR	Integer bitwise *or*

```
Scope (_SB) {
        Device (KBPD) {
                Name ( _ADR, 0×2000)
                Name ( _S0D, 0×0)
                Register (SystemIO, 8, 0, 0×0020, 8, TPLC)
                Register (SystemIO, 8, 0, 0×001A, 8, TPLR)
                External (H8_WRITE, MethodObj, IntObj, 3, IntObj, IntObj, IntObj)
                Method (_PS0 ) {
                        H8_RESET ()
                        H8_WRITE( TPLC, 1, 0×0F)
                        H8_WRITE( TPLR, 1, 0×04)
                }
        }
}
```

FIGURE 5.3   Sample of ASL script.

The features and format of the DSDT are retained to preserve the matured standard to describe the platform sufficiently. This makes it easier for users experienced with ACPI to use this framework. Any changes in hardware require only updating of the ASL file and restart of the subsystem.

The control methods to configure devices in most embedded systems are simply register reads and writes that incorporate only a small subset of ASL commands to realize subsystem implementation. Our implementation of ASL also supports fixed and hot-pluggable devices, multiple processors, on-chip peripherals, and USB and other communications buses.

5.3.3.1 Updating ASL Table to Accommodate New Device

The scope of a system device may be as a system bus, on-chip device, processor, or USB bus. These devices are the top-level hierarchical objects in a *Namespace*. If a device does not fall within the current scopes, a new scope object can be created with the new device object underneath it. All power information pertaining to a device must be described in the *ASL table (file)*.

The *ASL table (file)* component makes it easy to add a new device to the platform to make it accessible to a subsystem. As noted earlier, adding new hardware requires reloading the updated *ASL table (file)*. However, if a new device cannot be managed by existing control methods (*Methods*), the control methods must be incorporated into the *Methods* module and the changes require the subsystem to be recompiled.

5.3.4 ASL Translator

The *ASL translator* is a subsystem module that accepts the *ASL file* as input and converts it into a data structure in the native OS language that can be accessed by *power policy* to manage power on the platform.

The previous section described the ASL code as a XML-script plaintext and not a compiled code like AML. This removes the need for an interpreter and introduces a simple translator to map platform power information onto *Namespace*. It also improves portability as AML interpreters do not need to be developed for each platform. This module is similar to the *ACPI Table Management* that manages the tables in BIOS for ACPI. Simplifying the *ACPI tables* to a single DSDT table enabled us to replace an interpreter with a simple and flexible translator.

ASL translation is performed by two open-source tools; a lexical analyzer (*Lex*) and a parser (*Yacc*). The tools were initially used within compilers and are now incorporated into the RTOS-ACPI that translates the input *ASL file* into a C-structure called *Namespace*. When a subsystem intends to incorporate additional device information, it can update the *ASL file* which would correspondingly update the *Namespace* structure without changes to the subsystem.

5.3.5 Namespace

Namespace is a hierarchical structure that represents the power-managed devices in a subsystem. It is built by the *ASL translator* using the *ASL file* containing the differentiated system description table (DSDT) during subsystem initialization.

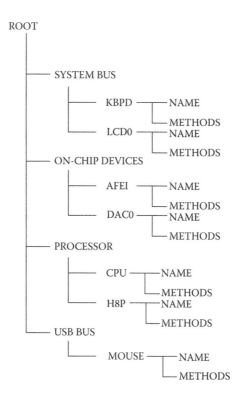

FIGURE 5.4 Outline of RTOS-ACPI namespace.

After *Namespace* is built, the subsystem updates the state of the device in *Namespace*. When a request for access is made, *Namespace* is accessed by the *Namespace manager* to find the control methods used to access the device.

Basically, *Namespace* organizes the named objects in the DSDT (data and control methods) in the form of a tree structure (Figure 5.4). It contains only the names—not actual functions, for example, a control method's name and register addresses would be present in *Namespace* but the methods that use these addresses reside in the *Methods* component. The relevant data (e.g., register addresses) would then be used by the control method.

5.3.6 NAMESPACE MANAGER

Namespace manager provides functionalities (shown in Table 5.4) to access and maintain the *Namespace* structure. Access is provided to the structure components via system calls that parse and evaluate named objects in the *Namespace*. There is no other means to access *Namespace*. *Namespace manager* has exclusive access. The responsibilities include device enumeration, obtaining and updating device information, and finding device control methods. In our implementation, *Namespace* is created by the *ASL translator* and not by *Namespace Manager*.

TABLE 5.4

Namespace Manager System Access Calls

System Call		Description
getData	*description*	Obtains data stored under object tag
	input	Device name, object tag (_DSS, _ADR)
	return value	Integer data stored in object tag
getMethod	*description*	Obtains detail of method to change device to specific power state
	input	Device name, power state
	return value	Returns structure that contains details of requested method
getTransTime	*description*	Retrieve transition time for state 1 → state 2
	input	Device name, state 1, state 2
	return value	Returns transition time (integer, milliseconds)
totalDevice	*description*	Obtains total number of devices under specific scope
	input	Scope tag (_SB, _USB) [*if NULL, then all devices in system*]
	return value	Total number of devices under scope

Each request to *Namespace manager* requires prior approval of the *basic policy manager* before accessing *Namespace* for the relevant control method. If the request violates any of the rules established in the *basic policy manager*, the operation is not executed.

5.3.7 METHODS

Most embedded platforms allow the device to be configured with register read and write mechanisms. The subsystem contains all the control methods to manage target platform power, thus removing the responsibility of managing device power from the device drivers.

The *Methods* default for controlling power is through hardware registers performing power state transitions. A subsystem can manage a corresponding device via its registers to configure the device to the required power state without the need to access *device driver*. The onus is on *Namespace manager* and *basic policy manager* to ensure that device access is always valid.

Certain embedded platforms have special methods to access power features. One example is the SH7727 T-Engine that uses an H8 co-processor as a power controller for a variety of on-board devices. Special methods are required to control the devices connected to this controller. Methods that send and receive commands over a serial interface (connected to the H8 co-processor) are required.

The original AML code would have contained all the methods to control these devices, but the design decision to remove the interpreter to reduce overhead necessitated this module. In addition, the ASL operator set was not capable of handling different control methods for each device. This module was created to contain all the methods or functions used by the devices in the platform. This brings hardware dependency into the subsystem, but is necessary for devices that cannot be controlled via simple register reads and writes. New methods can be incorporated into the source files of the subsystem.

5.3.8 Basic Policy Manager (BPM)

The *basic policy manager* is an important module that helps maintain the real-time constraints of the RTOS by validating all requests against its set of rules. This ensures that no request may require system power transitions that might violate these constraints. This module was not described in the ACPI specification, but was incorporated into our design to prevent any dangerous decisions by the *power policy algorithm* that might negatively impact the system (Figure 5.5).

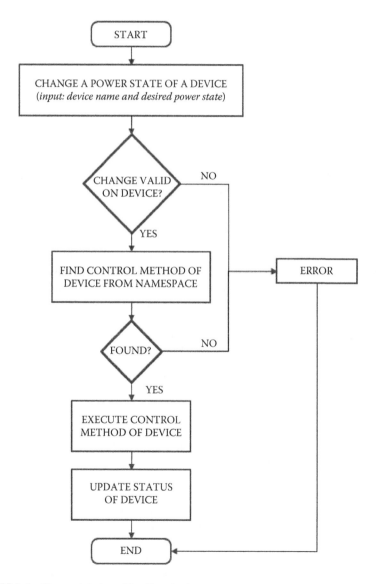

FIGURE 5.5   Power state transition flowchart.

More rules can be incorporated to help the RTOS. However, modifications should be kept simple to avoid significant overhead. Currently, this module supports two basic policies:

1. Transition of a device to low power (D3) shall not be permitted if the device has been opened by any application task for operation.
2. An error code is thrown when the advanced policy manager attempts to transit an unconnected device to any power state.

5.4 IMPLEMENTATION OF RTOS-ACPI FRAMEWORK

T-Kernel is a microkernel that implements only basic functions of a real-time operating system in the main body of the kernel but allows extended functions via middleware. We chose T-Kernel to implement our framework as the RTOS has a limited power management mechanism and requires a developer to define functions to handle power management. Figure 5.6 illustrates how we incorporated the framework into T-Kernel.

5.5 POWER MANAGEMENT POLICIES

The RTOS-ACPI subsystem alone is not sufficient to make power management comprehensive. For the framework to be complete, another important aspect to consider is the *power policies algorithm*. The performance of framework depends on an algorithm's ability to comprehend how the application will affect the devices on the target platform.

Making power management effective requires more than simply turning peripherals on or off. It is necessary for the framework to be flexible. Optimization can be achieved through techniques that coordinate granular sleep states instead of turning all peripherals on at once. In a power-managed system, the state of operation of each device is dynamically adapted to the required performance level to minimize power wasted by idle or underutilized devices.

Unfortunately, no policy performs well in all circumstances [7]. However, a policy can perform well with specific applications behavior. A monitor may be incorporated to track an application's behavior and use the most suitable policy.

5.5.1 IMPLEMENTING POWER POLICY ALGORITHMS

The interactions of *power policy algorithms* are representative of the method by which the RTOS directs power management decisions. This capability is within the proposed framework and is not part of a subsystem. The primary client of the subsystem is the *power policy algorithm* that introduces intelligence to the framework that utilizes RTOS-ACPI services to manage the corresponding device. The *power policy algorithm* calculates and makes power management decisions such as when to transit a device to a low power state and to what extent. The subsystem provides the *power policy algorithm* with the necessary information and services it may require to make an informed decision to manage system power.

An example is to use the *power policy algorithm* to predict device idle time [5] so that it might transition the device to a low power state. However, the algorithm

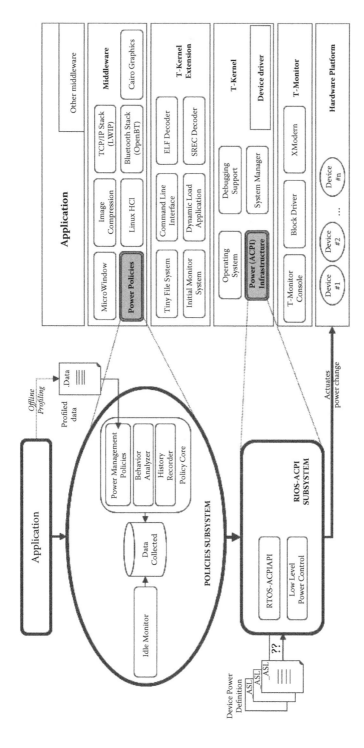

FIGURE 5.6 RTOS-ACPI framework in T-Kernel RTOS.

would use the device transition time and compare it with the estimated device idle time to decide whether this decision is profitable. The subsystem provides system calls (e.g., *DetStateTransTime*) to show the transition time of the device. These features assist the *power policy algorithm* to determine the merits of a policy decision. The subsystem also provides convenient services like putting all devices in a bus to a particular state, waking the entire system to the working power state, etc. A well-designed policy framework can gain from a subsystem leading to a synergic system.

The responsibility of a policy is to calculate and decide when best to change the state of a device. The validity of the decision should be left to the subsystem (*basic policy manager*) to determine. This would reduce the amount of device-specific data the policy needs to keep track. Device-specific data can also be stored in *Namespace* so that it can be accessed easily by different policies. While these device-specific data reside in *Namespace*, the policy is responsible for updating.

5.6 POWER SAVING AND REAL-TIME ABILITY

One aspect to be considered when developing policies for embedded system is balance between the amount of power saved and maintenance of real time. Another consideration is the extra power consumption required when the power manager decides to transition a device from busy to idle.

An algorithm must carefully consider overall power consumption when performing a state transition such as switching to a deeper sleeping state. The transition may incur a higher energy cost than the shallower sleeping state. A typical device is unable to switch immediately to another power state due to a state transition delay that utilizes energy. To compensate for this extra energy dissipation, the device must stay long enough in the next power state. Figure 5.7 illustrates state transitions.

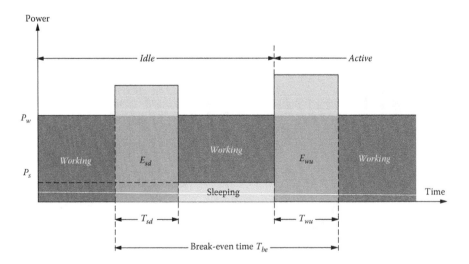

FIGURE 5.7 Power consumption graph illustrating break-even time.

P_w and P_s denote the power consumption in working state and sleep states, respectively. E_{sd} and E_{wu} represent the energy consumption for shutdown and wakeup. T_{sd} and T_{wu} are shutdown and wakeup delays. To save power, the idle time must be long enough to compensate for the extra energy E_{sd} and E_{wu}. Thus we can derive the following inequality:

The minimum length of idle time to achieve power saving is designated *break-even time* or T_{be}. If fail to consider the drawback of wakeup delay and see only the energy view, a system will shut down when idle period t is exceeds T_{be}. For the same idle time, a longer break-even time means less power saving due to overhead for recovering.

If a device receives a request from an application just as it is set to low power, it must wake up immediately. This triggers a delay penalty that may cause a serious performance failure in a real-time embedded system. The challenge is to find the right moment to wake the device to avoid the delay penalty.

5.7 CONCLUSION

We proposed a power management framework for a real-time operating system (RTOS) based on advanced configuration and power interface (ACPI) in this chapter. The comprehensive framework can be adapted to various RTOSs and does not require changes to applications or device drivers. The framework facilitates the rapid incorporation of application-specific power management strategies to optimize the power and energy consumption of embedded systems. The proposed framework can pave the way for standardizing the incorporation of power management with platform abstraction.

Future work aims to implement a set of power policy algorithms to exploit application behaviors for power management frameworks. This can be achieved by employing an application monitoring module to switch between policy algorithms in real time to adapt to the dynamic characteristics of a system.

REFERENCES

1. A. Grover. 2003. Modern system power management. *QUEUE*, October.
2. L. Benini et al. 2000. A survey of design techniques for system-level dynamic power management. *IEEE Transactions on VLSI*, 8, June.
3. S.K. Datta, C. Bonnet, and N. Nikaein. 2012. Android power management: current and future trends. *IEEE Transactions on ETSIoT*, 48–53.
4. HP, Intel, Microsoft, Phoenix, and Toshiba. n.d. Advanced Configuration and Power Interface Specification 4.0. http://acpi.info/spec.htm
5. C. Hwang and A. Wu. 1997. A predictive system shutdown method for energy saving of event-driven computation. In *Proceedings of International Conference on Computer-Aided Design*, 28–32.
6. Intel. n.d. ACPI Component Architecture Programmer Reference http://acpica.org/documentation/
7. Y.H. Lu, E.Y. Chung, T. Simunic et al. 2000. Quantitative comparison of power management algorithms. In *Proceedings of Design Automation and Test in Europe*.
8. Y.H. Lu, L. Benini, and G.D. Micheli. 2000. Operating system-directed power reduction. In *Proceedings of International Symposium on Low Power Electronic Design*.

9. E. Sheridan. n.d. Implementing Power Management on the Biscayne SH7760 Reference Platform Using the QNX Neutrino RTOS. www.qnx.com/developers/articles/article_296_2.html

10. WindRiver. n.d. VxWorks Platform 6.9. http://www.windriver.com/products/product-notes/vxworks6-product-note.pdf

6 Range of Benchmarks Required to Analyze Embedded Processors and Systems

Jeff Caldwell, Chris Fournier, Shay Gal-On,
Markus Levy, and Alexander Mintz

CONTENTS

6.1   INTRODUCTION

Since their introduction decades ago, embedded processors have undergone significant transformations. When embedded processor benchmarks started becoming popular early in the century, the embedded processors targeted were 8- and 16-bit microcontrollers or 32-bit microprocessors with little to no integration.

Today these processors have evolved into highly integrated devices. Even the low end of the computing spectrum is represented by 32-bit microcontrollers combined with large amounts of on-chip memory and sophisticated peripherals. Similarly, the high-end processors evolved into complex systems-on-chips (SoCs).

Associated with these transformations, the benchmarks used to quantify the capabilities of these processors have also grown in complexity. At the simplest level,

121

benchmarks such as CoreMark are used to analyze the fundamental processor cores. At the other end of the spectrum, complex benchmarks like DPIBench are used to analyze entire SoCs, system software stacks, and even physical interfaces.

This chapter explains the range of benchmark methodologies—from CoreMark to MultiBench to DPIBench—and guides readers in understanding appropriate application of embedded benchmarks.

6.2 COREMARK: BASIC APPLICATION-INDEPENDENT BENCHMARKING

At the simplest end of the processing spectrum, a need exists for benchmarks specifically targeted at processor cores. The Dhrystone benchmark has historically been used as a comparison tool for processors because it is free, small, easily portable, and displays a single number benchmark score. However, Dhrystone has several serious drawbacks:

1. Major portions are susceptible to a smart compiler that optimizes the work away and Dhrystone becomes more of a compiler benchmark than a hardware benchmark.
    a. This makes it very difficult to compare results when different compilers and/or flags are used.
    b. Since Dhrystone tries to create a specific mix of instructions, compiler optimizations change the mix and the result is no longer representative of the instruction mix for which the benchmark was designed.
2. Library calls are made within the timed portion and typically account for most of the time consumed by the benchmark. Since the library code is not part of the benchmark, it is difficult to compare results if different libraries are used.
3. The code is synthetic and does not mimic any behavior that can be expected in a real application.

The CoreMark benchmark ties a performance indicator to execution of simple code. However, instead of being entirely arbitrary and synthetic like Dhrystone, the benchmark code uses basic data structures and algorithms that are common in most embedded applications.

Furthermore, all computations are driven by runtime values to prevent code elimination during compile time optimization. This requires that every computation chain end in a result that is output to a volatile location such as printed to a screen (for example, by saving all results in a buffer and then printing the buffer after ending the test). Any step that is eliminated in the computation or control chain will lead to a different result. To exemplify these "unwanted" compiler optimizations, consider the line *while (cond) {stuff}*.

In CoreMark, *cond* is initialized and the initial value can be determined only at runtime. As a counter-example, Dhrystone contains a function in which *cond* is not initialized. A smart compiler can optimize the loop away since ANSI allows a compiler to choose any value for an uninitialized variable.

To be meaningful, *stuff* must end in a result output from the memory space of the benchmark (for example, by adding the end result of a computation to a memory location, then outputting it after the run). Dhrystone contains a function in which part of *stuff* does not modify the output of the benchmark. In other words, nothing else depends on the result; if that calculation were never done, nothing would be affected and a smart compiler could safely optimize it away.

To appreciate the workload value of CoreMark, it is worthwhile to dissect its composition. It is composed of operations on lists, 1-D arrays (strings and vectors), and 2-D arrays (matrixes).

6.2.1 LIST PROCESSING

Lists commonly exercise pointers and are also characterized by non-serial memory access patterns. In testing the core of a CPU, list processing predominantly checks how fast data can be used to scan through a list. For lists larger than the available cache of a CPU, list processing can also test the efficiency of cache and memory hierarchy.

CoreMark list processing consists of reversing, searching, or sorting a list according to the contents of its data items. In particular, each list item can contain a pre-computed value or a directive to invoke a specific algorithm with specific data to provide a value during sorting.

To verify correct operation, CoreMark performs a 16-bit cyclic redundancy check (CRC) based on the data contained in the elements of the list. Since CRC is also a commonly used function in embedded applications, this calculation is included in the timed portion of the CoreMark.

In many simple list implementations, programs allocate list items as needed with a call to *malloc*. However, in embedded systems with constrained memory, lists are commonly constrained to specific programmer-managed memory blocks. CoreMark uses the latter approach to avoid library calls to *malloc* and *free*. CoreMark partitions the available data space into two blocks:

```
typedef struct list_data_s {
      ee_s16 data16;
      ee_s16 idx;
} list_data;
```

One block contains the list (*list_head* pointers), and the other contains the data items (*list_data* items). This mimics common behavior in embedded designs in which data accumulate in a buffer (items) and pointers to the data are kept in lists or ring buffers. The *list_data* items are initialized at runtime from data that are not available at compile time.

Each *data16* item consists of two 8-bit parts, with the upper 8 bits containing the original value for the lower 8 bits. The data contained in the lower 8 bits are as follows:

0 to 2: Type of function to perform to calculate a value
3 to 6: Type of data for operation
7: Indicator for pre-computed or cached value

The benchmark code modifies the *data16* item during each iteration of the benchmark.

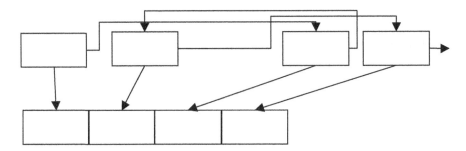

FIGURE 6.1   Basic structure of linked-list access mechanism. Each list item has a pointer to data and a pointer to the next list item.

The *idx* item maintains the original order of the list items so that CoreMark can recreate the original list without reinitializing it (a requirement for systems with low memory capacity). This need to recreate the list is another reason for storing the original value of the lower 8 *data16* bits in the upper 8 bits. The list is defined as follows; its structure is illustrated in Figure 6.1:

```
typedef struct list_head_s {
      struct list_head_s *next;
      struct list_data_s *info;
} list_head;
```

While it adds a level of indirection for accessing data, this structure is realistic and is used in many embedded applications for small to medium lists. The list is initialized on a block of memory that is passed to the initialization function so that malloc is not needed.

The linked list is initialized such that one quarter of the list pointers point to sequential areas in memory and three quarters of the list pointers are distributed in a non-sequential manner. This is done to emulate a linked list on which items added and/or removed have disrupted the order of the original list, after which a series of additions have come from sequential memory locations.

The list head is modified during each iteration of the benchmark, and the next pointers are modified when the list is sorted or reversed. At subsequent iterations of the benchmark, the algorithm sorts the list according to the information in the *data16* member, performs a test, and then recreates the original list by sorting back to the original order and rewriting the list data.

Since pointers on CPUs can range from 8 to 64 bits, the number of items initialized for the list is calculated so that the list will contain the same number of items regardless of pointer size. In other words, a CPU with 8-bit pointers will use one quarter of the memory that a 32-bit CPU uses to hold the list headers.

6.2.2   MATRIX PROCESSING

Many algorithms use matrices and arrays, resulting in significant efforts to optimize this type of processing. These algorithms test the efficiency of tight loop operations along with the ability of the CPU and associated tool chain to use ISA accelerators such as MAC units and SIMD instructions. The algorithms are composed of tight loops that iterate over the whole matrix.

CoreMark performs simple operations on the input matrixes, including multiplication by a constant, a vector, or another matrix. CoreMark also tests operations on part of the data in the matrix by extracting bits from each matrix item for manipulation. To validate that all operations have been performed, CoreMark again computes a CRC on the results from the matrix test.

Within the matrix algorithm for CoreMark, the available data space is split into three portions: an output matrix (with a 32-bit value in each cell) and two input matrixes (with 16-bit values in each cell). The input matrixes are initialized at runtime based on input values that are not available at compile time. During each benchmark iteration, the input matrixes are changed based on input values that cannot be computed at compile time. The input matrixes are then recreated with the last operation, and the same function can be invoked to repeat exactly the same processing.

6.2.3 STATE MACHINE PROCESSING

An important function of a CPU core is the ability to handle control statements other than loops. A state machine based on *switch* or *if* statements is an ideal candidate for testing that capability. The two common methods used to implement state machines are (1) using switch statements and (2) using a state transition table. Because CoreMark already utilizes the latter method in the list processing algorithm to test load and store behavior, it uses the former method (*switch* and *if* statements) to exercise the CPU control structure.

Figure 6.2 illustrates the simple state machine used in the tests. It has nine states. The input is a stream of bytes initialized to ensure that all available states are visited based on an input that is not available at compile time. The state machine tests an input string to detect whether the input is a number; if it is not a number, the machine will reach the invalid state. The entire input buffer is scanned by the state machine.

To validate operation, CoreMark keeps count of how many times each state was visited.

During each iteration of CoreMark, some of the data are corrupted based on input that is not available at compile time. At the end of processing, the data are restored, also based on inputs not available at compile time.

6.2.4 CoreMark PROFILING AND RESULTS

Since CoreMark contains multiple algorithms, it is interesting to demonstrate how its behavior changes over time. For example, looking at the percentage of control code executed (samples taken at each 1000 cycles) and branch mispredictions in Figure 6.3, it is obvious where the matrix algorithm is called. This is portrayed by the low misprediction rate and high percent of control operations indicative of tight loops (for example, between points 330 and 390).

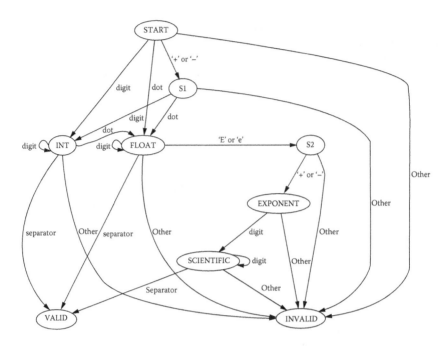

FIGURE 6.2 Overall functionality of CoreMark's state machine processing.

By default, CoreMark requires the allocation of only 2 Kbytes of memory to accommodate all data. This minimal memory size is necessary to support operation on the smallest microcontrollers so that it can truly serve as a standard performance metric for any CPU core.

Figure 6.4 examines the memory access pattern during benchmark execution. The information is represented as a percentage of memory operations that access memory within a certain distance from the previous access. It is easy to see that the distance peaks are caused by switching between the algorithms (since each algorithm operates on a slice of one third of the total available data space).

Figure 6.5 shows selected CoreMark results that display interesting patterns. The fact that some results depend on the compiler version and flags makes clear that these details must be included; otherwise a useful comparison is impossible. Therefore the run and reporting rules for CoreMark require that exact tool versions and compilation flags used be reported along with any performance results.

A. Blackfin results (1,2) show a 10% increase in performance when moving from GCC 4.1.2 to GCC 4.3.3, a reasonable expectation for a newer compiler version.
B. Results (8,9) show an even more pronounced difference of 18% for a mature compiler, while other results (10,11) (12,13) show minor effects only, as all those compilers are based on the GCC4 series.
C. The compiler can also balance code size against performance as we can see in (3,4). The compiler and platform are the same, but when directed

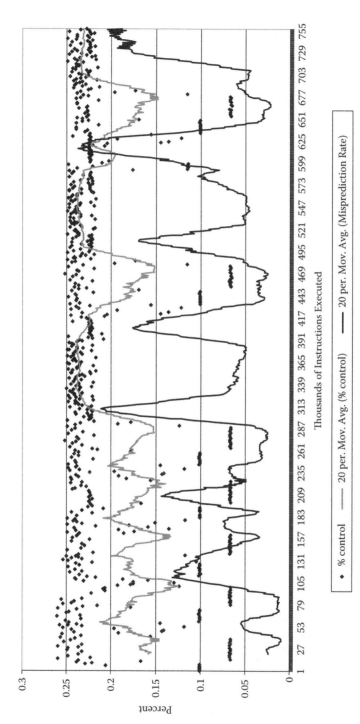

FIGURE 6.3 Distribution of control instructions and mispredictions with CoreMark execution. The graph illustrates exact data points and a moving average for control operation to emphasize the extremes.

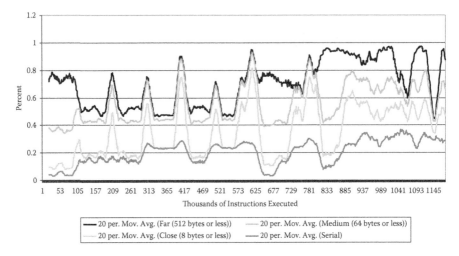

FIGURE 6.4 Distribution of memory access distance over time during CoreMark execution. About 20% of the time, memory access is serial, with peaks of non-serial access likely due to state machine operation. This access pattern will test the cache mechanism (if any) and memory access efficiency of systems without cache.

Processor	Compiler	CoreMark/MHz
1 Analog Devices BF536 0.3 393 MHz	gcc4.1.2	1.01
2 Analog Devices BF536 0.3 393 MHz	gcc4.3.3	1.12
3 Microchip PIC24FJ64GA004 32MHz	gcc4.0.3 (dsPIC30, Microchip v3_20)	0.93
4 Microchip PIC24FJ64GA004 32MHz	gcc4.0.3 (dsPIC30, Microchip v3_20)	0.75
5 Microchip PIC24HJ128GP202 40MHz	gcc4.0.3 (dsPIC30, Microchip v3.12)	1.86
6 Microchip PIC24HJ128GP202 40MHz	gcc4.0.3 (dsPIC30, Microchip v3.12)	1.29
7 Microchip PIC32MX360F512I (MIPS32 M4K) 72MHz	gcc3.4.4 MPLAB C32 v1.00-20071024	1.71
8 Microchip PIC32MX360F512I (MIPS32 M4K) 72MHz	gcc3.4.4 MPLAB C32 v1.00-20071024	1.90
9 Microchip PIC32MX360F512I (MIPS32 M4K) 80MHz	gcc4.3.2 (Sourcery G++ Lite 4.3-81)	2.30
10 NXP LPC1114 48MHz	Keil ARMcc v4.0.0.524	1.06
11 NXP LPC1114 48MHz	gcc 4.3.3 (Code Red)	0.98
12 NXP LPC1768 100MHz	armcc 4.0	1.75
13 NXP LPC1768 72MHz	Keil ARMCC V4.0.0.524	1.76
14 Texas Instruments OMAP3530 500MHz	gcc4.3.3	2.42
15 Texas Instruments OMAP3530 600MHz	gcc4.3.3	2.19
16 TI Stellaris LM3S9B96 Cortex M3 50MHz	Keil ARMCC V4.0.0.524	1.92
17 TI Stellaris LM3S9B96 Cortex M3 80MHz	Keil ARMCC V4.0.0.524	1.60
18 Xilinx MicroBlaze v7.20.d in Spartan XC3S700A FPGA, 3-stage pipeline, 2K/2K cache 62.5MHz	gcc4.1.1 (Xilinx MicroBlaze)	1.48
19 Xilinx MicroBlaze v7.20.d in Spartan XC3S700A FPGA, 5-stage pipeline, 4K/4K cache, integer divider, 62.5MHz	gcc4.1.1 (Xilinx MicroBlaze)	1.66

FIGURE 6.5 CoreMark/MHz performance data. Complete benchmark environment details are available at www.coremark.org.

to build a smaller executable (Os –mpa), performance drops 19% versus optimizing the code with –O3. The distinction is even sharper with results (5,6) at 30% performance difference (using –mpa switch). Note: compiler switch information is available from CoreMark website reports.

D. Other compiler options affect how much the compiler tries to optimize the code. Results (7,8) show a typical effect from safest (–O2) versus aggressive (–O3) optimizations of about 10%.

E. When operating frequency is scaled up, the system memory and/or on-chip flash cannot always maintain a 1:1 ratio with CPU speed. It is common to see extra wait states on the flash at higher processor frequencies. When the code resides in flash, the efficiency (expressed as CoreMark/MHz) is impacted; (14,15) shows an efficiency drop of almost 10%. For (16,17) the wait state effect is even more pronounced as the CPU:memory frequency ratio can be maintained 1:1 up to only 50 MHz; when operating at the highest device frequency (80 MHz), the ratio drops to 1:2, resulting in an efficiency drop of 15%. However, running at 80 MHz still yields an absolute performance improvement of 25% versus running at 50 MHz.

F. Results (18,19) explore the situation in which the cache is too small to contain all the data. In 18, the cache is exactly 2 Kbytes, which will fit all the data just barely, but leave no room for function arguments that must be passed on the stack. This causes a small amount of bus traffic external to the cache, but when the cache is enlarged (19), performance improves 10% despite the change to a less-efficient five-stage pipeline versus a three-stage pipeline for the first implementation.

6.3 MULTICORE BENCHMARKING

Benchmarking and performance expectations become more complicated for multicore systems. The multicore technology era is solidly upon us and most processor vendors offer such products. However, system developers realize that adopting multicore technology can present as many challenges as benefits.

One of those challenges lies in analyzing the potential performance of a multicore processor including multiple processors on a SoC. Installing multiple execution cores into a single processor (and continuing to increase clock frequency) does not guarantee greater multiples of processing power. Furthermore, depending on the application, a user has no assurance that a multicore processor will deliver a dramatic increase in system throughput.

Despite this caveat, the right combination of processor and programming techniques can scale well with the number of cores. Success will depend on the parallelism available in the algorithm and the amount of independence between the calculations performed.

The effect of parallelism can be demonstrated in a case study generated by AMD and Zircon Computing [2]. This case study illustrates the common situation in which legacy applications written for conventional single-core processors cannot easily leverage the additional cores in multicore or multiprocessor systems without substantial modifications. In other words, it reveals how the single-threaded embedded benchmark representative of a typical legacy networking application can be converted to a scalable multithreaded program.

The case study begins by noting that a suite of networking benchmarks exhibits characteristics that make it difficult to improve performance using conventional HPC software platforms (i.e., MPI, OpenMP, or CUDA). Three characteristics contribute to this limitation. The first is the presence of both independent and dependent computations. This arises because the loop computations in some EEMBC networking benchmarks (such as the QoS benchmark) support multiple threads that can execute multiple iterations simultaneously; with other EEMBC benchmarks, the outcome

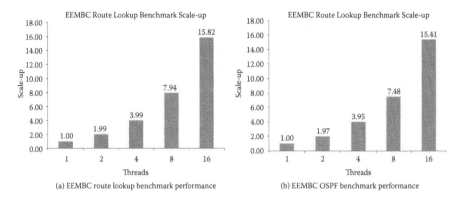

(a) EEMBC route lookup benchmark performance

(b) EEMBC OSPF benchmark performance

FIGURE 6.6 Multicore benchmark performance scaling.

of an iteration depends on the outcome of a previous iteration, making it difficult to distribute and run loop iterations simultaneously across multiple threads.

The second characteristic is the presence of computations with ultra-short execution times. The impact is that the potential thread context switching overhead exceeds the time needed to run a single computation in a single thread. The third characteristic is the presence of computations that access global data. This increases contention as multiple threads compete for the same data.

After applying the task and data parallelism methods depicted in [2], benchmark performance results were generated using 1, 2, 4, 8, and 16 threads and compared with the performance of the baseline sequential application implementation. The authors choose 16 threads as their maximum thread count since it matched the 16 available processing cores in the AMD hardware and maximized performance by disseminating work across the processing resources without oversubscribing the system.

The charts in Figure 6.6 showing only two benchmarks indicate a nearly linear performance improvement as the number of threads increases. The situation with a real-world program can be considerably more complicated, however, especially when the number of contexts exceeds the resources available.

6.3.1 Memory Contention and Other Bottlenecks

Assume that a program is composed of a varying number of threads (it is reasonable to have hundreds of threads in a relatively complex program). If the number of threads exactly matched the number of processor cores, it would be possible for performance to scale linearly, as the benchmark above shows. However, the number of threads typically exceeds the number of cores and performance depends on other factors such as memory, I/O bandwidth, intercore communication, OS scheduling support, and synchronization efficiency.

The memory bandwidth of a multicore processor depends on the memory subsystem design that in turn depends on the underlying multicore architecture. *Multicore* implies a shared or distributed memory architecture. Shared memory, typically associated with homogeneous multicore systems, is accessed through a bus and controlled by a locking mechanism to avoid simultaneous access of the same memory by multiple cores.

The shared memory structure provides a straightforward programming model as each processor can directly access memory (Figure 6.7). However, memory access can become a bottleneck when too many cores try to access a shared memory simultaneously. A bottleneck also indicates that the memory architecture does not scale well with increasing numbers of cores.

Unless an application is running on 'bare metal' (directly on processor hardware without operating system support), OS scheduling will also play a big role in determining multicore implementation behavior. Scheduling determines how processes are assigned priorities in a priority queue and is also determined by availability of on-chip processing resources (based partly on the OS's ability to monitor availability of hardware resources such as cores or hyperthreads).

Multicore benchmarks must consider all these issues, but this is more easily said than done. Just as there has never been a simple way to measure the performance of a "normal" single-core processor and reduce it to a single measure of "goodness,"

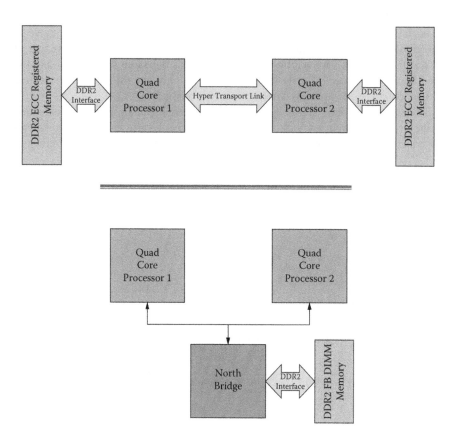

FIGURE 6.7 Two dual-core, shared memory architectures with different memory subsystems. Both approaches present performance advantages and disadvantages.

it is exponentially more difficult to measure the performance of a multicore device and produce a single figure of merit.

One of the first things EEMBC learned while working with multicore benchmarks is that the combined interactions of all the factors outlined above result in marked performance differences, even among similar platforms. Tests on two dual-core processors, for example, showed very different rates of speed-up depending on the number of concurrent streams and specific benchmarks run (Figure 6.8). From this information, one can tailor software to align with the benchmark characteristics that yielded the highest performance on the specific processor.

For example, using the results below and known benchmark characteristics for the workloads, a user could implement an image rotation algorithm to operate on two independent images in one system. On another system, he or she might choose to have both cores cooperate to speed processing of one image at a time. From a parallelism perspective, a multicore benchmark must target two fundamental areas of concurrency: data throughput and computational throughput. Benchmarks that analyze data throughput show how well a solution can scale over scalable data inputs. It can be useful to increase the number of data inputs to determine the point at which performance begins to degrade.

In developing such a benchmark test, one challenge is that the code must be thread-safe to support simultaneous execution by multiple threads. In particular, it must satisfy the need for multiple threads to access the same shared data and the need for a shared piece of data to be accessed by only one thread at a given time without compromising required performance throughput.

To demonstrate computational throughput, the approach above can be extended further by developing tests that can initiate more than one task at a time, implementing concurrency over both data and code. This will demonstrate the scalability of

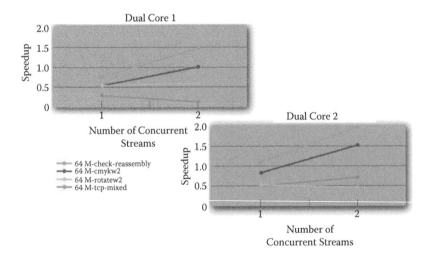

FIGURE 6.8 Comparing two dual-core platforms demonstrates how results can vary and depend on multiple factors.

a solution for general-purpose processing. As a benchmark, this requires synchronization between the contexts and a method for determining when the benchmark completes.

Data decomposition is where an algorithm is divided into multiple threads that work on a common data set, demonstrating support for fine-grained parallelism. These threads are distributed based on the number of available processor cores. Efficient processing is possible because the cores within the multicore device are closely distributed and can support high bandwidth, low latency data transfers.

6.4   MULTIBENCH BENCHMARK SUITE

EEMBC implemented an extensive suite of multicore benchmarks called MultiBench. It utilizes an API abstraction to make it easy to support SMP architectures. In the embedded industry featuring such a wide variety of architectures (even considering only SMP), flexibility is key. Hence, when porting to a new platform, only 13 API calls are needed to allow the MultiBench framework (multi-instance text harness or MITH) and all the benchmarks to run and take advantage of parallel execution.

Furthermore, many systems support the POSIX threads (Pthreads) interface and the abstraction was chosen carefully to ensure direct mapping to Pthreads. This means that if the system already supports Pthreads, no porting is necessary.

The multicore benchmarks are delivered as a set of workloads; and a workload consists of one or more work items. While it can be laborious to analyze an application and benchmark workloads for commonality, users can pick from this list and select the workloads that most closely resemble their applications. The official release of MultiBench 1.0 contains 36 workloads that use common embedded algorithms in a way that enables multiple cores to enhance performance. When running these workloads, the user may change the number of work items running in parallel and the affinity of each instance of the work item to take advantage of available computing resources in the target platform.

Unfortunately, it is not possible to represent the capabilities of a multicore platform with a single number so EEMBC derived three numbers or marks that best reflect the multicore processor's throughput and performance scaling. The MultiMark consolidates the best throughput of workloads having only one work item that uses only one worker. The ParallelMark consolidates the best throughput of workloads having only one work item that uses multiple workers. The MixMark consolidates the best throughput of workloads having multiple different work items. These workloads are closest to the workloads on actual systems. Each mark is further based on two figures of merit derived from each workload:

Performance factor—This concept depicts the best throughput of the platform, defined in iterations per second. Each platform will execute a workload a specific number of times per second. The best configuration is platform-dependent.

Scaling factor—This concept defines how well performance scales when more computing resources are brought to bear on the workload. Among several ways of utilizing computing resources in parallel, MultiBench 1.0 workloads with the MultiBench framework to test most of them (functional decomposition is the lone exception and will be addressed in the next revision of MultiBench). By limiting the available resources to execute only one work item at a time and then comparing the throughput to the best throughput for the workload, we gain an insight into how well a platform scales for a workload.

To calculate the scaling factor, first calculate the geometric mean of the throughput for the workloads with only one work item at a time enabled. The geometric mean is used instead of the arithmetic mean because it does a better job of de-emphasizing outlying high and low values and yields a more representative average. The next step is dividing the performance factor by that number.

Sample scores on a simulated platform with 16 cores (Figure 6.9) are interesting even without considering the details of specific workloads. For example, consider the fact that the MultiMark scaling factor is 8.9 rather than a number closer to 16 that might be expected with 16 cores. This factor strongly hints that the system can use only about half the computing resources available to it for any particular problem.

This is likely related to memory bottlenecks since single-worker workloads show little synchronization and thus are less dependent on the synchronization efficiency of the platform and operating system. A more detailed examination of the individual results may be needed to yield more answers and highlight the trends better than a single-number representation can.

Figure 6.10 compares the results from two dual-core platforms. Note the significant difference in performance factors of the two platforms. While both use Linux and GCC and have the same core frequencies, they are based on different architectures and have different memory hierarchies. Platform #1 has a shared L2 cache while Platform #2 has separate L2 caches.

Basic analysis of the results show that platform #1 is 30% faster on MultiMark and 80% faster on ParallelMark. Understanding the performance differences does not necessarily require sophisticated performance analysis tools. Often, simply reviewing the specific scores of different workloads and correlating those with

	Performance Factor	Scale Factor
MultiMark	10.9	8.9
ParallelMark	10.5	4.7
MixMark	4.5	8.8

FIGURE 6.9 Sample scores on simulated 16-core platform.

	Platform #1		Platform #2	
	Performance Factor	Scale Factor	Performance Factor	Scale Factor
MultiMark	33.3	1.9	24.7	1.8
ParallelMark	24.8	1.9	13.8	1.7
MixMark	11.8	1.7	7.2	1.5

FIGURE 6.10 Sample scores from two dual-core platforms running at 2 GHz.

architectural differences can be enough to indicate where the differences arise. For example, synchronization overhead and cache coherency traffic both result in significantly lower ParallelMark values for Platform #2.

The combined effect of using all these workloads provides a comprehensive view of multicore processor behavior. Some results derived by using data throughput tests can serve as examples of the methods described above.

6.4.1 MultiBench Results

A look at how two different quad-core x86 processors perform on a benchmark suite can illustrate these subtleties. Both chips are connected to 4 GB of 667-MHz DDR2 memory. The test results shown on the following charts were generated by the MultiMark (SWM), ParallelMark (MWM), and MixMark (MIM) benchmarks.

Figure 6.11 illustrates how a "Brand X" processor performed on three different MultiBench tests as workloads increased. The horizontal scale indicates that the workload increases from one context (at left of the chart) up to 20 contexts (at right). The vertical axes on this and subsequent charts have been normalized so that the performance of a single context is always 1.0. This makes it easier to see how performance scales—or does not scale—with increasing workloads.

The performance throughput of the Brand X quad-core processor increased as the number of workloads increased. If the performance did not increase, there would be no reason to use a multicore processor in this application. However, the performance did not increase linearly. The maximum performance with 20 contexts is just short of three times the baseline performance with one context.

Even with four processor cores working on 20 tasks, overall performance throughput triples (which many would consider a reasonable increase). However, performance on the multithreaded MIM test is lackluster, maxing out at less than 2.0 times the baseline performance. On the other hand, at least performance does not decrease significantly with increasing workloads (although it does decrease slightly with the MWM test).

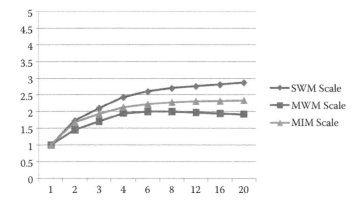

FIGURE 6.11   Brand X x86 quad-core processor with single memory subsystem. The Y-axis is relative performance compared to a single context; the X-axis is the number of contexts.

Figure 6.12 shows results of the exact same tests conducted on a nearly identical system, but with a competing Brand Y processor. Unlike the previous example, the results graph shows a pronounced "kink." This processor's performance on the MWM benchmark increased linearly up to four contexts (one per processor core) and then declined as more contexts were added. Its performance on the SWM and MIM benchmarks was somewhat more intuitive. Performance gradually increased, then plateaued around 8 to 12 contexts. The kink may have been caused by a variety of factors, including memory bus overload, synchronization overhead between contexts, and L2 cache bottleneck.

Figure 6.13 shows results for a system with two Brand X processor chips, each with four cores, for a total of eight processor cores. As in the first test, the system has 4 GB of DDR2 memory but, in this case, the two processors share it. All the memory is local to one of the processors and the other processor accesses it via a shared

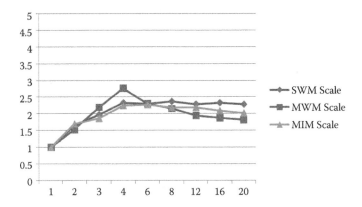

FIGURE 6.12 Brand Y x86 quad-core processor with a single memory subsystem. The Y-axis is relative performance compared to a single context; the X-axis is the number of contexts.

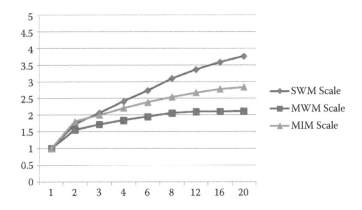

FIGURE 6.13 Dual Brand X x86 quad-core processors sharing single memory subsystem. The Y-axis is relative performance compared to a single context; the X-axis is the number of contexts.

link between the two chips. This gives one of the processors a built-in advantage, although both can access all available memory.

The data show that performance scales better than it did with the single-processor (four-core) system from Figure 6.11. Peak performance on SWM is about 3.75 times the baseline. While nowhere near eight times the performance, it represents a substantial improvement. The two multicore benchmarks (MWM and MIM) also show steady growth, and, in fact may have grown further had the workload been increased.

6.5 APPLICATION-SPECIFIC BENCHMARKING AND BENCHMARK CHARACTERIZATION

The CoreMark and MultiBench benchmarks focus solely on processors using very generic workloads. The results may or may not reflect the characteristics of specific application areas. If a user has a need to evaluate an architecture for suitability for a specific application, a set of benchmarks targeting those applications may make sense. The question is, how relevant will a given benchmark be for a target application, since each application will deviate somewhat from any benchmarks?

Although benchmark characterization is a challenging exercise, it can be used to explain the actual workload presented by a benchmark. Benchmark characterization is the process of finding a set of unique characteristics to classify benchmarks. These characteristics must have two properties: (1) they should be predictive of performance on a wide variety of platforms, and (2) they should allow benchmark consumers to find good proxies for their own applications in the benchmark set. The goal is finding just the right characteristics so that the required hardware capacities of the benchmarks are known. In this regard, benchmark characterization can be used to develop new benchmarks and classify existing ones to determine whether they stress a system in similar or different ways compared to the applications of interest.

In one example, Conte et al. [1] utilized the EEMBC application benchmarks to devise a method of describing application behavior independent of the underlying platform. These EEMBC benchmarks (AutoBench, ConsumerBench, DENBench, NetBench, OABench, and TeleBench) represent a wide variety of workloads and provide industry standard tools for understanding processor performance.

In this project, the authors collected benchmark characteristics for the MIPS, PowerPC, and x86 architectures. Characterization was primarily achieved through trace-driven simulation, collecting data from cache design experiments, and the distribution of dynamic instructions. Functional unit requirements were gathered by simulating an idealized superscalar machine. To remove all effects other than true dependencies, the idealized machine assumed perfect branch prediction, a perfect cache, and infinite pipeline width.

Kiviat graphs are useful for representing the metrics of simulations, making it easier to visualize multivariable data in a way that reveals program behavior. They reflect hardware-dependent, hardware-independent, and hardware design characteristics that together generate an architecture-independent characterization of each benchmark.

Figure 6.14 shows an example Kiviat graph that represents the results of automotive benchmark characterization. It depicts relative usage of simple integer arithmetic instructions (IALU), logic instructions (LSU), integer multiplication and division

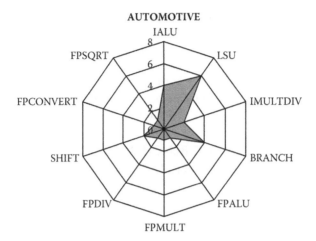

FIGURE 6.14 Kiviat graph showing function unit usage for automotive benchmarks (85% utilization).

(IMULTDIV), branch instructions, floating point ALU, floating point multiplication, floating point division, shift, floating point conversion, and floating point square root instructions. While the axes give a quantitative picture of each variable, the shape gives a qualitative picture easily compared to other Kiviat graphs.

Figure 6.15 represents the average characteristics from all other benchmarks in the series and demonstrates the great variety among benchmark suites. This is significant because it reveals the application-specific nature of each benchmark suite and shows that more than one suite should be run to demonstrate the capabilities of a processing platform. It also highlights the fact that a benchmark such as CoreMark is useful but is clearly not sufficient to determine the capabilities of a processing platform.

To show how to interpret data in Kiviat graphs, note that the automotive suite exhibits similar loading on functional units as the Networking v2.0 suite, with the difference that Networking v2.0 has slightly higher requirements for branch instructions. The figure analysis shows the workload varieties among the various benchmark suites, but further analysis also shows significant variety within specific workload categories, the details of which can be found in Reference [1].

6.6 TRANSITIONING TO SYSTEM LEVEL: SCENARIO BENCHMARKS

The demand for the types of benchmarks described to this point continues to grow and it is also crucial to consider application-specific system-level standard (scenario) benchmarks. Because embedded processors have evolved into complete systems through integration on SoCs, it is not possible to evaluate their performance thoroughly using portable C code.

Scenario benchmarks take into account more system-level features and can be viewed as 'black-box' benchmarks. In general terms, they must specify a benchmark's input, expected output, and interface points. In other words, what is inside a benchmark does not matter as long as it can handle the input and deliver the expected output.

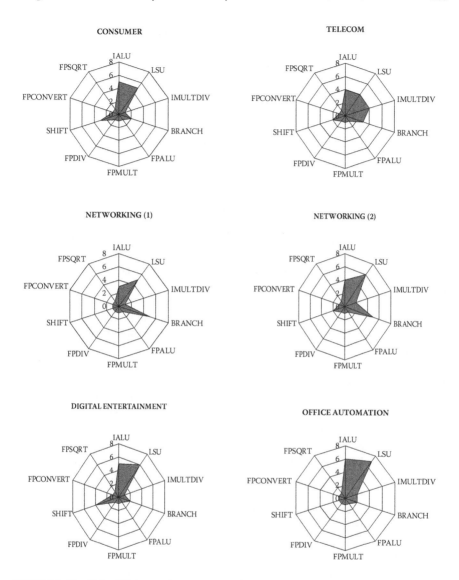

FIGURE 6.15   Using Kiviat graphs to represent functional unit distribution (85% utilization).

Although scenario benchmarks represent advancement for SoC testing, they create a challenge in the analysis of the results. While an embedded platform is the sum of its components, running a scenario benchmark makes isolation of any specific component a difficult task. The scenario benchmark requires hardware (SoC, memory system, I/O) and software (operating system, application, runtime stacks). Finding bottlenecks can be as difficult as agreeing on a performance metric.

To look at this in more detail, consider a benchmark for a deep packet inspection (DPI) appliance. In this application, operations are carried out on packet headers at all seven layers of the open systems interconnect protocol stack, performing stateful inspection, deep packet inspection, intrusion detection and prevention, anti-virus, anti-malware, anti-spam, and content filtering. Some DPI systems also implement firewalls on the basis of application or the authentications of individual users. Such a task list often requires dedicated RISC and/or DSP processors operating at both the control and data planes.

The DPI appliance represents an optimal case of why modern computer system architects talk about strict separation between the control plane where management of compute operations takes place and the data plane where real-time operations are performed on streams of data entering or leaving a computer appliance. When packets are operated on at the network edge, it is critical to look at the benchmarks for management and administration of a network element and, more importantly, the ability to examine and manipulate packets at wire speed. These latter functions take place in the data path through data plane operations. Appliances are expected to perform deep packet inspections in which the actual contents of a data packet are probed to characterize the traffic and identify potentially malicious content.

The most common DPI appliance architecture has one or more datapath processors operating directly on packet traffic at wire speed, along with a centralized control plane processor that manages the operations of the packet engines that offload packet header inspection. In some DPI architectures, the Ethernet or T1/E1 processors that control transceivers on physical or data link planes also perform some rudimentary traffic management functions on behalf of the datapath processor. Encryption and decryption engines may be separate chips or may be integrated into network processors.

Benchmarking such functions, however, has taken a long time to catch up to capabilities. While the functions of DPI share commonality of definition not seen at the turn of the millennium, the specific implementations allow a wide disparity in how such functions are measured. To date, no common method can handle performance testing and validation of DPI throughput. Such a testing approach must consider the various threat vectors used in attempting to transfer infectious payloads into a network.

Without a common standard by which to compare performance across all these variables, consumers of DPI technologies lack an objective means of selecting a solution from the myriad vendor offerings available and are often at a severe disadvantage when attempting to select the most suitable solution to protect their information systems. Throughput numbers from various DPI system vendors have been known to be off as much as 90% from the numbers claimed on their data sheets and collateral marketing materials.

Benchmarking DPI systems presents a further challenge: performance may vary even between two systems utilizing the exact same chips, depending on which features the vendor implemented. Furthermore, in the absence of a complete system, network processor performance can only be estimated. For this reason, when defining and developing a benchmark for DPI systems (DPIBench), EEMBC focused on the testing of complete systems.For a networking application like this, the significant metrics for system performance are throughput, latency, and quantification of the number of flows. Although efficacy is also an extremely important metric, the analysis of the efficacy of next-generation firewalls and DPI appliances is subject to too many issues due to the variety of methods for tracking efficacy.

Furthermore, for throughput and capacity, performance standards must come from the network side of the interface because the relevant viruses and malware a DPI appliance confronts can change daily, if not several times a day. A standard efficacy benchmark thus would have little relevance outside a single day's virus portfolio. However, a measure of packet analysis at a given network speed would remain relatively constant even over changing content-filtering loads.

DPIBench focuses on the network traffic side of a DPI appliance utilizing test equipment from companies such as Breaking Point Systems, Ixia, and Mu Dynamics. The most important common protocols to analyze for DPIBench were judged to be HTTP, HTTPS, SMTP, P2P, and FTP. It is possible that higher layer presentation and application protocols could be added as sites to move to more general use of XML and related languages. The secure version of HTTP is included deliberately, as many attacks now use encrypted methods and secure transport protocols to disguise sources of infections.

It is also important not to focus solely on the protocols in packet transport and consider which protocols are assigned to which ports. Port spoofing and the use of non-standard ports must be carefully monitored if a test methodology is to exhibit validity in this context.

For an industry standard benchmark, the identification of viruses represents a difficult area in which to achieve any cross-platform standardization because many software and DPI appliance companies treat their virus signature databases as proprietary intellectual property. Many such databases employ different formats, and updates to databases may occur at different times via different update methods.

The WildList.org site represents a partial standardization of common virus signatures. For the purposes of DPIBench, the most recent suite of common signatures published at http://www.wildilst.org/WildList is utilized. One must assume, however, that 20 to 30% of the entire content of WildList changes on a monthly basis.

In its methodology, EEMBC runs two tests: one test emphasizes performance and in the other viruses are injected at random points. The latter test ensures that the correct number of ports is terminated and that any traffic using non-standard ports is identified. We identified the following variables as relevant to the test:

- **Traffic types**, identified as I or G where I represents IPS/IDP data transfers and G represents gateway anti-virus and anti-spyware file transfers
- **Data model**, identified as m or t where m represents minutes of maximum continuous traffic under IPS/IDP and t represents file transfers for anti-virus and anti-spyware
- **Ports and protocols**, identified as s or n where s represents tests that run a common protocol on standard ports and n represents tests running a common protocol on altered ports
- **Encryption and compression**, identified as u, z, or e where u represents unencrypted traffic, z represents testing the transfer of compressed files, and e represents the testing of traffic over HTTPS.
- **Efficacy sanity check**, identified as c or v where c represents tests without malware and v represents tests that include viruses identified by WildList.org.

The assumption in developing such tests is that a full test suite can be run in 30 minutes or less. Tests are dynamic in format, particularly due to the rapidly changing nature of virus signatures identified by WildList. It is important to run a clean test without viruses as a baseline, followed by tests with random virus insertion.

The concern that someone could optimize a system for a benchmark may be minimized by constantly expanding the tests for the number of viruses detected and updating the tests for newer viruses on WildList. Demonstrating benchmarks across various virus lists shows the variance potential of a DPI appliance, but it raises a question unique to security benchmarks: can the industry handle less than 100% repeatability of a benchmark because of constant virus updates?

In a typical setup, the system under test will be connected to a test device capable of generating traffic matching the conditions of the designated IPS, IDP, and gateway test scenarios. A system under test will act as a router that transfers traffic from source to destination IP address, with both addresses residing on the test device.

6.7   CONCLUSION

In summary, benchmarking embedded systems involves far too many variables to allow a single number or even a single benchmark to provide all the answers. The processor matters as do the number of processors, the way in which multiple processors are connected, and an overall system configuration. And, of course, the operating system, tools used, and intended application matter. No single answer is correct for every use.

For this reason, benchmarks like CoreMark serve to isolate basic processors. MultiBench addresses multicore systems, and application-specific and scenario benchmarks help assess the suitability of a system for a specific problem. Careful benchmark construction helps thwart those who may want to "game" a system in order to misrepresent their products. Where possible, third party certification adds another layer of credibility to a set of benchmark results. While benchmarks are not panaceas and do not absolve a system engineer of the due diligence required for selecting or creating a system, they can point an engineer in directions that will be most promising.

REFERENCES

1. T.M. Conte and W.W. Hwu. 1991. Benchmark characterization. *IEEE Computers*, Jan., 48–56.
2. A. Mintz, D.C. Schmidt, J. Balasubramanian et al. n.d. Using Zircon Technology to Optimize the Performance of the EEMBC Suite on Embedded Eight-Core AMD Opteron6134 Processors. www.zircomp.com.

7 Networking Embedded Systems: An Exciting Future*

Roberto Saracco

CONTENTS

7.1 HAVE WE REACHED THE END OF THE ROAD?

The evolution we have witnessed in the past fifty years in electronics, optics, smart materials, biotechnology, and in all the fields using these technologies has been relentless. Although we see no signs of having reached a plateau, we know that a physical limit to progress lies somewhere. While the ceiling appeared to be approaching in many fields like electronics, engineers found ways to circumvent. However, that does not change the reality that physical limitations exist.

In economics, we have seen what happens when we reach a ceiling such as running out of liquidity: the downward spiral of stock markets in the second half of 2008 is a clear statement of the havoc that occurs when progress is suddenly stopped.

---

* This chapter is a clone of one I wrote five years ago for a book titled *Internet Networks* edited by Kris Iniewski for CRC Press. The editor of this book requested permission to re-use that chapter, and I agreed with the provision that I could rewrite it to make it timely. I was really surprised to see how much of what I wrote five years ago is still valid. Some experiments became realities and new developments appeared, but the main trends are still there. Hence, it took very little time to transform "will be" into "is" in many place and add new material here and there. I hope I can say the same five years from now. For readers wishing to follow the pace of innovation: http://blog.telecomfuturecentre.it

The technology evolution has progressed with such regularity that it no longer surprises us. We are accustomed to innovation. Actually we have built a world that relies on it. If technological evolution stopped next year, we would need to reinvent the way we do business and that would cause tremendous problems.

Looking at the physical barriers like the speed of light, the quantum of energy, the smallest dimension that exists, we can determine where the ultimate limit lies. The good news is that such limits are very far from where we are today. As Richard Feynman explained in a lecture: "There is plenty of room at the bottom!" At the present pace of evolution we won't reach it for the next few centuries. This does not mean, however, that such limits will ever be reached. Actually, I feel that we will discover unsolvable issues long before we reach those physical barriers.

The investment required for chip production plants is growing exponentially and returns on investments require huge revenues. This means pushing toward huge production volumes, with prices for individual products decreasing in order to reach the widest possible market. The economics is already slowing down the creation of new plants, and new production processes may circumvent what we see as upper boundaries today.

Energy is becoming a bottleneck to evolution (Figure 7.1). China since 2010 consumes more power than the United States but its per capita consumption is a fraction of the consumption of the U.S. We can expect more than 8 billion people to populate the Earth before 2020. In terms of energy, the increase in consumption will not be 25 to 80%. The increase will likely reach 800% because the increased population will consume on average the power now consumed on average by a U.S. citizen because their lives will be improved by technology.

We simply do not have the energy available at a sustainable cost that would be required if these assumptions were met. This means that either global power consumption will not be on the average of the consumption of a U.S. citizen in 2008 or that we will have found ways to dramatically decrease our power consumption and increase energy production—both will be required to maintain the huge population.

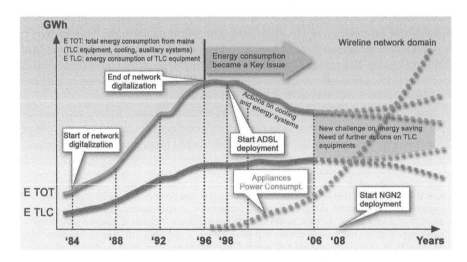

FIGURE 7.1   Growth of energy consumption.

The energy availability and cost will influence worldwide technology evolution over the next decade. The shift toward a "greener" world, although vital, will increase the impact of energy on technology. The bright side is that the energy "crunch" will force investment in alternative energy sources and measures for decreasing consumption. This, rather than slowing evolution, is likely to shift the direction of evolution, accelerating the deployment of optical networks that are much more energy economical than copper networks. This will mean radio coverage by smaller cells since the energy required to cover a surface decreases (approximately) with the square of the number of cells used.

At present and for the horizon that is reasonable to consider today (not beyond 2050), we can be confident that technology evolution will continue at a different pace in various sectors as has been the case in the past fifty years. Overall, it will continue at a rate similar to what we have experienced in recent decades. Saying that the pace will be substantially unchanged should not deceive us into believing that "business as usual" will continue for three reasons.

First, Moore's law claims that today's performance will double in the next eighteen months. That means that in eighteen months we will cover the same distance that took us forty-five years to reach. That is quite a change!

Second, some of today's technology is reaching a plateau. Silicon is an example. Pushing beyond today's progress would require so much investment that evolution would be economically unsustainable. However, these levelling-off technologies will be first flanked and then replaced by new technologies that will continue the evolution of performance. For example, carbon and new computing paradigms will overtake silicon and ensure the validity of Moore's law in the third and fourth decades of this century.

Third, performance increase exerts a linear effect on an ecosystem until a certain threshold is reached. Beyond that point, the dynamic is seen as a change of rules rather than a performance increase. Think about electronic watches. In the 1970s, electronic watches were very expensive. As prices dropped, more people could afford them. At a certain point, the cost dropped almost to zero and the market for the watches disappeared. The industry had to reposition itself into the fashion industry. Long gone are watch advertisements selling precision. The marketing value of a Swiss-certified chronograph in terms of precision dropped to zero.

A similar progression is occurring with the deployment of broadband networks and their enabling optical infrastructures. After a bandwidth of 1 Gbps is achieved (and possibly earlier), it will become impossible to market increased bandwidth at a premium. Bandwidth value will drop to zero and marketers will need to find new features.

Notice how the thresholds link technology with market value. These two factors are changing the business model (rules of the game) and disruption takes place. As this happens, consolidated industries need to reinvent themselves. New industries find leverage to displace the incumbents.

In the discussion on technology evolution we will reach this point over and over. The issue to consider is not whether technology is reaching its evolution limit, but whether it will lead to disruption thresholds.

7.2 GLOBAL INNOVATION

Innovation used to be easier to predict because a few companies in a few countries led the way. The evolution of infrastructures was so easy to predict that the International Telecommunications Union (an international standardization body based in Geneva) published an annual table detailing the status of telecommunications infrastructures and service penetration. Maps showed the progress made and projected penetration ten and twenty years in the future. As an example, it would take a country nineteen years to move from 1 to 10% telecommunications penetration but only twelve years to move from 10 to 20%, eight from 20 to 30%, and so on.

The advent of wireless changed the world. India formerly had less than 3% telecommunication density (and that is still the percent for fixed lines). However, it progressed from 0 wireless density to 20% in ten years. In July 2012, India cell phone users totalled 920 million, reaching 79% penetration. China wireless penetration increased in ten years from 0 to over 30% and exceeded 1 billion cell phones early in 2012 (density of 75%). Not only big countries are leaping ahead. Vietnam by late January 2012 had over 118 million mobile phones—a density of 134%!

In the infrastructure domain where globalization has decreased prices, a broader market indirectly affects local situations. As a result, the service domain has a distribution cost that is basically zero. After an application is developed, it can be made available over a network with no distribution cost hampering its marketing. The network is also playing another trick: an application that makes sense in the U.S. market can be developed and marketed from India. Innovation is no longer confined to the domains of a few rich companies or countries because of huge investment barriers. The real barriers have moved from money availability to education availability.

Increases in educating engineers in India and China—not their huge potential markets—have placed these countries at the forefront of innovation. The market of innovation is now global.

This globalization of innovation will continue through the next decade. It is not by chance that the U.S. is pursuing the goal of increasing scientific education and the numbers of engineers as the way to remain at the forefront of scientific and economic progress.

Optical and wireless networks will further shrink the world. Distance is already irrelevant for information flow and is becoming irrelevant for delivery of a growing number of products and services. Surely it is becoming irrelevant for innovation.

Offshoring will become more practiced by big industries and smaller companies will inshore innovation from anywhere and profit from their abilities to localize a global world. Inshoring will continue. Offshoring will be viable only until a labor cost differential will "plague" the world and will eventually disappear. Politics, regulations, and cultures will be the determining factors in the evolution. From a technological view, the Earth will be no bigger than a small village.

The Web 2.0 paradigm is already evolving from a network of services and applications made available by a plethora of (small) enterprises and huge investment of a few (big) enterprises to become Web 3.0 in which interactions among services and applications will serve specific users' contexts. A question like "when is the next train?" will become answerable because the Web will understand my context and know that I am looking for information about a train to Milan. This type of advance is not a small step.

Again, progress is a matter of "glocalization." We are moving from the syntax, from infrastructures providing physical connectivity, to semantics, to the appreciation of who I am, and this includes the understanding of who I was (the set of experiences shaping my context), and the forecasting of who I will be (the motivations and drives to act). This may seem Orwellian, frightening, and definitely not the way to go.

However, the evolution over a long period must be beneficial; otherwise it will not become entrenched. We can expect that the balance between what is technologically possible and what the market is buying will depend on individuals and their membership in communities. Contextualization can raise many issues—privacy, ownership, democracy, and the establishment of new communities continuously reshaping themselves. Contextualization is not likely to result from an intelligent, Orwellian network, but rather from the increased intelligence of my devices, and that is under my control. I will make the decision, most of the time unconsciously, of what to share of my context and the network will be there to enable it.

My devices (the "my" is the crucial part) will act as an autonomous system, absorbing information from the local environment and thanks to the network, from the global environment. Devices will let me communicate with my context, my information, my experiences, my environment and, of course, my friends and acquaintances in the same seamless way as today when I walk into a room and act according to my aims, expectations, and other factors.

Therefore, telepresence, one of the holy grails of communications, will also be "glocal." I will communicate locally and remotely as easily as if they involved the same technology. This will be possible because technology will continue to evolve. Although the list of technologies to consider will be very lengthy, we should take a look at a few of them in terms of the evolution of functionalities that will become available.

7.3 DIGITAL STORAGE

Digital storage capacity has increased by leaps and bounds over the last fifty years (Figure 7.2). The original solutions to digital storage (magnetic cores, drums, tapes) have disappeared and replaced by new technologies such as magnetic disks, solid state memory, and polymer memories (on the near horizon).

Hard drives (devices using magnetic disks for storage) reached 744 Gb per square inch in 2012 and are expected to reach 1,880 Gb per square inch by 2016. This corresponds to 30 to 60 TB for a 3.5-inch HDD and 10 to 20 TB for a 2.5-inch HDD; 100 TB capacity in the residential market will become common by the end of the this decade. The leap in magnetic storage density was achieved through heat-assisted magnetic recording (HAMR). The use of helium-filled HDD (expected in 2103) will reduce power consumption by over 40% per TB.

Solid state memories have advanced significantly. Compact flash cards are now inexpensive and ubiquitous. They were invented in 1994 and advanced from 4 MB to 512 GB in 2013. Solid state disks (SSDs) based on flash technology appeared in

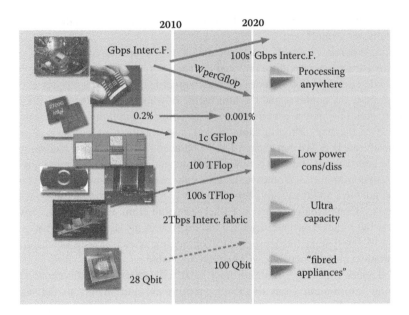

FIGURE 7.2 Evolution of digital storage.

2007. In April 2012, Intel announced the 910 Ramsdale generation with a capacity of 800 GB using 25 nm technology, a read speed of 2 GBps and a write speed of 1 GBps.

This increase of capacity sets the SSD on a collision route with magnetic disks in certain application areas like Mp3 players and portable computers. SSDs consume only 5% of the energy required by magnetic disks and are shock-resistant up to 2000 G (corresponding to a 10-foot drop). The bit transfer rate has already increased significantly and exceeds that of magnetic HDDs.

Polymer memories have seen an increased effort by several companies to bring them to the market. They are now used in some applications (like toys) that require low capacity. These memories are made by printing circuit components on plastic and are precursors of fully fledged printed electronics. Their advantages over other types of memories are extremely low cost and potentially huge capacity. An area the size of a credit card can store several terabytes of data.

Data will be stored both at the edges and in networks. In 2008, Ericsson foresaw a 1 TB cell phone to be available in 2012. What we saw in 2012 was flash cards in cell phones that reached 128 GB and levelled out, mostly because of the parallel development of SSD and the cloud. SSD is used in lap tops and tablets; the cloud is a way to extend storage for mobile devices. The cloud trend will continue in this and the next decade fuelled by faster and faster mobile networks. Access to the cloud is swift and effective, and it sparked the economic interest of some players (Apple, Google, plus a few telecoms) to lock in customers by becoming their "data managers."

Home media centers will manage the total life activities of a family through multi-TB storage with tens of TBs expected within this decade as normal in the home environment. EB (a billion billions of bytes) storage will become common for data warehousing by data-based companies like Google, Snapfish, Flicker, Facebook, and companies that will replace them in the next decade. Institutions and governments will harvest on a day-by-day basis the digital shadows of their constituencies to offer better service. Their data storage will grow into tens of PBs.

Raw data generated by sensors will be transformed into economic value through statistical data analysis approaches and correlation (big data). Storage is becoming one of the most important enablers for businesses in this decade. What are the consequences of this continuous increase of storage capacity? Clearly, we can store more and more data and information. However, the real point is that this huge capacity is changing the paradigms and rules of the game, affecting the values of networks and impacting their architectures.

Since data are everywhere, the flow of data will no longer be restricted from the network toward the edges. The other direction will be as important as that one. In addition, we are going to see the emergence of local data exchanges, edges to edges, and terminal to terminal. The first evolution makes the uplink capacity as important as the downlink (leading to the decommission of ADSL and its replacement by vectoring and fiber). The second evolution emphasizes the importance of transaction-oriented traffic. Updates will achieve greater importance and possibly be perceived as the real values delivered by some providers. Raw data will make sense only if they can be converted into perceptible chunks of information relevant to a user (person and machine).

As we will discuss in the next section, storage may disappear from sight, replaced by small "valuets" (mixtures of applications, sensors, and displays representing meanings of value to a user). We see this already in tiny apps on iPhones that mask data, information, transactions required, and even the applications used. The storage cards that embed communications and applications provide hints of future advances.

7.4 PROCESSING

Processing is still evolving to provide more and better performance. New chip architectures contain multicores that support parallel computations, thus multiplying capacity without increasing processing time and resulting power requirements. Three-dimensional architectures support the layering of cores. In principle, we have the technology to pack hundreds of cores within a single chip. However, increasing capacity by multiplying cores does not mean a genuine increase in capacity because not all computations can take place in parallel and more cores require more parallelism. This is why biotechnology is progressing faster than Moore's law: sequencing a genome takes full advantage of parallelism, data crunching usually does not.

However, processing evolution is no longer alone on the axis of increased performances. Other factors like reduced energy consumption and ease of packaging are gaining importance. As in the past, continuous increases in processing performance expanded the market. Decreasing energy consumption and cheaper ways to package chips are dramatically opening new markets.

Intel noted in 2005 its intention to achieve a 100 times reduction of energy per GFLOP by 2010 and met its target. Programs now under discussion by several innovators aim for "zero power consumption." That is obviously impossible and clearly refers to decreasing power consumption to a point where scavenging energy from ambient sources will be sufficient to power the devices.

A decrease in power consumption enables the packaging of more processing power in hand-held devices like cell phones. However, the concern is not the resulting reduced battery drain; it is potential danger. A 500-watt cell phone will burn a hand long before it drains its battery.

A second concern may have far-reaching consequences. Very low consuming devices may be powered via alternative sources. Examples are the conversion of sugar in the human blood stream into energy to power tiny devices delivering drugs and monitoring body chemistry, exploiting temperature differentials to generate electrical power, conversion of surface vibrations into energy for sensors placed in road surfaces to measure traffic, and conversion of wireless radio waves into energy using evanescent waves to power sensors in closed environments.

As the cost of producing sensors decreases, the economic issues shift to operation and power sources. Progress in decreasing power consumption has made us confident that within twenty years we will see an explosion of sensors to match the explosion of data. By 2005, only a tiny fraction of microprocessors were installed in computers. Most of them were used in microwave ovens, remote controls, automobiles, and electronic locks just to name a few. In this decade, most devices already embed microprocessors and most have the capability of connection in and to networks. This advance will change dramatically the way we perceive and use objects.

Part of this change will be enabled by "printed electronics" manufactured from a derivative of ink-jet printing. The process is inexpensive—one-third to one-half the cost of silicon etching now used for chips. Printed electronics devices are cheaper to design (again, a third the cost of etching silicon) and can embed processing, storage, a radio antenna and, if needed, a touch-based interface. They also eliminate the cost of packaging. In principle it will be possible to write on devices as easily as we now stick labels on them.

This evolution affects microprocessing. We will continue to see changes in the opposite direction—the development of "super crunchers." Processing speed will increase incrementally via massive parallel computing systems utilizing hundreds of thousands chips per machine. Speed will exceed the PFLOPS of today (100 PFLOPS as of 2012), and the EFLOPS in the next decade (billions of billions of floating point operations per second) will also see more diffused use of the cloud computing paradigm in the business environment and in business-to-consumer activities. Consumers are unlikely to appreciate what occurs behind the scenes. They will be unaware that some of the services they use are achieved by massive processing from a cloud computing infrastructure.

Looking further into the future, we can speculate that cell phones and other wireless devices may form a sort of cloud computing system for resolving interference issues, thus effectively multiplying their uses and efficiency. The major hurdle on this path (already demonstrated as technically feasible from an algorithmic view) is the energy required by the computations and communications to be performed by the devices. The energy requirements make such systems impossible today and for the coming years.

7.5   SENSORS

Sensors are evolving rapidly, becoming less expensive and more flexible. They embed communications functions and thus are ready to form local networks. Sensors represent a "Pandora's box" of services. Think about the thousands of applications of the iTouch and iPhone (and to a more limited extent on tablets) that exploit accelerometer sensors.

Drug companies are studying new ways to detect proteins and other substances. Research in the past that required long and expensive tests executed by large and very expensive machines can now be performed less expensively, faster, and easier by sensors alone or in combination.

Nanotechnology is finding new ways to create sensors with specific sensing capabilities. The sensors can detect single molecules. For example, a single carbon nanotube can be "drugged" by a specific molecule and serve as a sensor that can detect a certain substance and block the flow of electrons.

Some of these sensors will be embedded in cell phones. One phone sensor can analyze breath as a person talks and assess the presence of markers for lung cancer over time. Markers for early detection of other diseases are under development. SD-like cards containing tens, and soon hundreds, of substances will be plugged into cell phones and allow the detection of a variety of illnesses well before clinical signs appear.

The drive to miniaturize sensors, to make them more flexible and responsive to the environment, will continue. Hundreds of sensors will be constantly producing meaningful data and enhancing communications as we will see in the discussion on statistical data analyses.

Other researchers are investigating e-textiles. These special fibers can be woven into clothing to sense a variety of conditions and presence of special substances like sugar and proteins and thus provide the data to detect several pathological conditions. Printed electronics will help cut production costs and sensors will be deployed in almost any object. Pick up an item and it will be ready to interact.

Sensors are also providing the technology to transform a collection of objects into an environment. Context awareness will make significant advances because of sensor presence everywhere.

Intel announced a new research program at the end of 2008: its wireless identification and sensing platform (WISP) that can identify any object, including a body. The company has released WISP to a research team to support applications development. The device is basically a miniaturized radiofrequency identification (RFID) system forming a continuous interconnected fabric.

The present RFID technology will transform into active components with sensing capabilities as prices of sensors decrease. This is not expected to happen before the end of this decade. In the meantime, more objects will embed sensors and some will be used for identification, thus eliminating the need for RFIDs. However, RFID may be utilized in new services, for example, monitoring activities of a person and creating a digital signature of his whereabouts. This may be helpful for detecting changes in routine of elderly people that may indicate problems.

The transformation of an object into an entity that can become aware of its environment will change business practices. A producer will be able to remain in touch with the user of a product and ultimately transform the product into a service. This will enable new business models and require transformations of producing organizations. Most producers will not be prepared for this change but will find it difficult to resist because the competition will be ready to exploit the marketing advantages provided by these new context-aware objects. Some producers turn over their product communications and on-board flexibility to third parties to let them further increase the features and resulting perceived values of products. This openness in turn will give rise to a variety of architectures, making network platforms and service platforms true service factory and delivery points.

Sensor research will create ripples in today's established dogmas like the ubiquity of information processing (IP). Energy efficiency considerations are driving sensor networks to use non-IP communications and many more sensor networks will use ad hoc protocols instead of local and backbone networks now using IP. Identity and authentication will expand to cover objects; this may lead to new approaches for assessing identity. The SIM card is very effective for identification and authentication, but it has not satisfied the banking systems and may not represent the future of identification. In fact, cell phones equipped with sensors detecting biometric parameters may provide even better authentication mechanisms and separate the terminal from the user; that would appease the banking systems.

Finally, the needs for self-standing sensors and reducing sensor energy consumption are pushing researchers to seek ever better autonomous system theories and applications. Advances in this area will exert profound effects on network ownership and management architectures since autonomous systems eliminate the need for a central control to deliver end-to-end quality and will undermine the foundations of modern telecom operators.

7.6 DISPLAYS

Display technology brought us the very popular wide flat screens and utilized screens in growing numbers of devices, from digital cameras to cell phones (Figure 7.3). Digital frames invaded our homes via many electronic devices. Some dreams have not come to pass, for example, the holographic screens that were supposed to occupy our living rooms according to futurists in the 1960s. Many available technologies are bound to progress even further, particularly in the direction of price drops to end users. Production process improvements represent the single most important factor in this progress.

The lower cost makes it possible to use screens everywhere in line with our perception of a world based on visual communications. The telephone is a very successful compromise. The telephone created such a strong new communication paradigm that most people prefer talking to video communication which is now considered far more intrusive. Telephone communication brings a user close to other parties, but the younger generation is on the way to massively adopt video

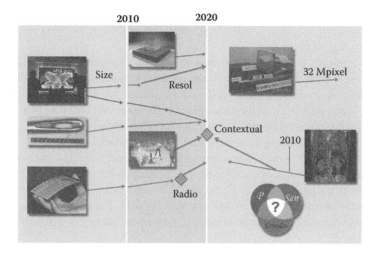

FIGURE 7.3 Display evolution.

communications. However, other areas of progress are important because of their perception impacts.

The resolution of the human eye is approximately 8 Mpixel. The brain composes signals received from the eyes into a bigger window whose resolution is about 12 Mpixel. Present HD television screens have 2 Mpixel resolution achieved using 6 Mdots (a triplet of red, green, and blue constitutes one resolution pixel). Hence, although we are awed by the quality of the images, our brains are not fooled. We are looking at a screen not at reality. We watch a show but we are not "at the show."

Japanese electronics industries set a goal of a 32 Mpixel screen (and the required production chain) by the end of the this decade. A few 4K screens reaching 8 Mpixel thresholds are already available. A person looking straight at one of these screens cannot differentiate the view from reality. We already utilize 8Mpixel resolution in consumer electronics. Most digital cameras have at least that kind of resolution. The Nikon D800 reflex camera announced in 2012 exceeds 32 Mpixel resolution.

However, people seldom look straight at anything. We may not be aware that our eyes constantly scan the environment. This scanning activity allows the brain to create a larger image and convey the feeling of "being there." To replicate this sensation, we must confine the scanning activity to the screen "real estate," that is, we must be sufficiently close to the screen and the screen dimensions must create an angle with our eyes exceeding 160 degrees. When we look straight at a screen, the angle captured by the eyes is slightly smaller than 130 degrees. The increasing dimensions of entertainment screens and their increases resolution will lead us into make-believe situations before the end of this decade. The bandwidth required to transmit that amount of information exceeds 100 Mbps. Only optical fiber connection will achieve that (at least in the distribution network; the last meter or so can be wireless).

Although we will have LTE, LTE+, and FRA, a better device that can handle those kinds of speeds will always be on the horizon. However, it does not make economic sense to chew all available spectrums for these types of services.

The home situation is different. An optical fiber may well terminate into a gateway that will beam information wirelessly at speeds up to 1 Gbps, possibly using bands in the 10 to 50 GHz range.

Smart materials for displaying images and clips will become more available. We already have a special varnish that can change its colors to create images. Electronic ink can display black and white text and Mirasol technology can display color images. We can expect a significant progress in this area that will lead by the end of this decade to ubiquitous display capabilities for many objects.

This progress will change the look and feel of products and, as noted for sensors, is bound to change the relationships of producers and users. Furthermore, these capabilities coupled with open systems and with open service creation platforms will enable third parties to provide services for any object. Displays are the ideal interfaces for humans because we are oriented visually. The coupling of touch sensors, gesture recognition, and other kinds of "intention" detectors will pave the way to new services.

The underlying assumption is that objects will be connected to networks directly or more often through local ambient networks. The connections in many instances will be based on radio waves although strong competitors may be utility power lines within a given area. The fibered telecommunications infrastructure is likely to stop at the entrance of a structure on the assumption that fewer wires are better.

7.7 STATISTICAL DATA ANALYSES: "BIG DATA"

The quantities and varieties of data becoming available through networks are growing exponentially. Cell phones are equipped with a growing number of sensors. The data generated by 5 to 7 billion cell phones concerning location, temperature, and movement (direction and speed) will create an avalanche. Special sensors may provide even more information. Telecom operators will have to reach an agreement on how to handle these data and make them available in a neutral way to preserve privacy.

These data can be used to monitor traffic in urban areas, detect the onset of epidemics, plan and monitor public transportation, develop precise maps of environment pollution, and examine social networks and interest generated by an advertisement or event. They may be used during emergencies or for urban planning. They can be used by shop managers to dynamically change their shop windows.

Clearly, data derived from cell phones represent only a fraction of the total data harvested by sensors and potentially available for analysis. Many objects will embed sensors and most will be connected to networks. Thus, the data they gather may be made available to third parties. Homes, department stores, schools, hospitals, parks, cars—any entity can generate data and make them available. The possibilities are endless.

We are learning how to analyze massive data banks. This type of analysis will develop further in the coming decade to include the analyses of distributed data banks, leading to the generation of metadata. Public access to metadata will stimulate the development of many services and new industries will be created. The step from raw data to metadata is crucial. Metadata should be able to capture what is of interest in a set of data, masking in a secure way any detail that can be used to trace the owner or generator of data used to create the metadata. The absolute guarantee of a decoupling of data and metadata is a prerequisite to allow the publication of metadata.

It is likely that most metadata will remain invisible to the public and will be transformed into useful information via services. We are already seeing this trend in several applications available on the iPhone. As an example, a user seeks a weather forecast for a certain area. The information is conveyed to the user without revealing the source. A service available at the touch of a button may actually integrate data from several weather forecasts based on a success track.

These data will be distributed to several databases and their harvesting and exploitation will generate traffic with varied characteristics. On the harvesting side, traffic will likely be in the form of billions of tiny transactions. The usage form will probably consist of bulk data transfers feeding statistical data analyzers in the forms of images and video clips to end users.

Wireless and optical networks will constitute the supporting infrastructure. Cloud computing may offer the computational capabilities required for data analyses in several cases and support for data sharing. Enterprises will generate ideas to leverage potential data value. The ideas will not require massive investment in computation structures if they are made available on demand through networks. Note how a significant portion of these enterprise-generated services may take the form of applets residing on users' terminals such as cell phones. After activation, they may generate traffic and computation by a network.

In summary, the availability of massive quantities of data is changing the approach to data analyses from algorithmic processing to statistical data processing. This change implies access to distributed data banks and significant computation capabilities that will be better satisfied by cloud computing ("better" refers to economic cost). The communication fabric is the key enabler at the levels of raw data harvest and processing into metadata balanced against specific application, customer demand, and distribution of results.

The huge availability of data, within enterprises and in the public domain, may be addressed by a relatively new discipline known as Big Data—a domain addressing the correlation of huge data streams. YouTube has a huge data set, but the correlation among the data is not massive. It lists movies, directors, actors, and other entertainment details that may be managed via classical database architecture. Similarly, telecom operators that have millions of customers and manage billions of call data records have huge data sets but have not correlated them effectively to date. Thus, we cannot say telecom operators are managing Big Data. However, the correlation of their data records will create interesting information and require new computational approaches that will fall into the Big Data classification. Big Data are characterized by three Vs:

Volume: only huge quantities constitute Big Data.
Variety: data sources should exhibit some heterogeneity so that correlation can provide added value.
Velocity: data should change with a certain frequency, otherwise classical processing approaches may be better.

In this decade, we will see a tremendous evolution in data correlation technology and in our ability to derive understanding from this correlation. The evolution will affect many areas such as logistics, sales, healthcare and, of course, science. Telecommunications will be at the core of the Big Data evolution as providers of huge amounts of data, as transporters, and also as managers of data and metadata sets.

7.8 AUTONOMIC SYSTEMS

The growth of independent network providers, particularly at the edges of the old networks, led to a new paradigm for management and exploitation. In addition, the "dumb terminals" connecting at the edges of the network have become very sophisticated devices. Their intelligence levels are comparable to and may exceed the intelligence of a network. These devices no longer terminate communications and services arriving from the network.

They manage their own local networks of relationships with other devices, sometimes acting as intelligent gateways that use networks as tunnels for pure connectivity. The intelligence has moved to the edges of the network, creating a need for a different paradigm to understand, create, operate services. The paradigm of autonomic systems extends well beyond the client–server arrangement that assumed some sort of local intelligence in a terminal.

The basic sequence of telecom engineering has always been planning, designing, engineering, and control, control, and control. All measurements made within a network are intended to provide information for control. The more complex the network, the more sophisticated the controls. The wide dissemination of computers in the 1980s allowed fully centralized control and telecom operators built their own control centers. As more equipment and technology found its way into networks, engineers and researchers developed standards to ensure the centralized control of the new diversity. Those in the field remember CMIP and SNMP—two systems for controlling network elements based on the premise that a single comprehensive view of a network was required to ensure fault control, maintenance, fair use of resources.

The advent of Internet consisting of thousands of interconnected networks owned by different entities was viewed with suspicion by many telecom engineers because the quality of its service was based on best efforts of individual networks. As Internet traffic grew, many predicted its imminent collapse because of a lack of centralized management.

Although local segments have experienced and will continue to experience outages and service interruptions, traffic continues to grow and no collapse has occurred. The catastrophic events of September 11, 2001, created havoc in the telecommunication infrastructure but left the Internet unscathed. The only way to communicate with people in New York on that day was through Internet messaging.

The Internet is not a typical network in the sense that it uses different wires. Its wires are shared with the telecommunications network. One reason the Internet works so well is its use of one of the most reliable infrastructures on the planet: the telecommunication infrastructure. The Internet is resistant to local faults because it is not subject to a hierarchy of control. Control is local and distributed. A malfunction at point A will not affect communications transiting through A because many equally useful alternatives allow traffic to bypass A. Internet is not an autonomic system in a full sense. The components within its boundaries act as autonomic systems for routing information.

A popular example of an autonomic system is the Roomba vacuum cleaner that became available about ten years ago. The Roomba is basically a robot with a goal: vacuuming the spaces around it as far as it can reach. The spaces are cluttered with obstacles like furniture and people (advertisements claim that the Roomba can avoid running over a cat although a cat will take appropriate action by stepping away). Over time, the Roomba learns how the space is structured and devises the best vacuuming strategy. If the space changes by the addition or reconfiguration of furniture, the robot will change its strategy accordingly.

The home environment is an ideal laboratory for testing autonomic systems. Actually, a home can become an autonomic systems if all its devices communicate and cooperate with one another. As new devices are installed, they make themselves known and become part of the environment. As the external environment changes (connections are updated or replaced), the autonomic home will react and reconfigure itself.

The evolution of wireless in the direction of more local areas will exploit the technology of autonomic systems. Devices will be surrounded by wireless clouds. A device will check overlaps with other clouds and try to establish connection at a semantic level. It will identify itself, communicate its characteristics, objectives, and needs, and expect to receive similar information from other devices sharing the cloud. This exchange of information may be followed by a "handshake" to bind the devices into a single system. Through continuous sensing of the environment, the system will evolve to accommodate changes of participants and behaviors.

7.9 NEW NETWORKING PARADIGMS

The advent of autonomic systems, the multiplication of networks, the huge storage capacities created at the edges of networks (terminals, cell phones, and media centers), and the growing intelligence in outside networks will significantly change networking operations. Efforts for thirty years focused on exploiting the progressive penetration of computers to make networks more intelligent. A simple economic drive drove the evolution: a network is a central resource whose cost can be split among users. It makes sense to invest in an intelligent network to provide better services at low cost.

The first dramatic shift was triggered by cell phones and mobile networks. If you wanted to develop a network and decided to use a fixed line network to provide services, you would have to invest almost 100% in the system. However, if you delivered the same services using a mobile network approach, the network cost would be 30% and the terminal cost (likely to be paid by customers) would be 70%. As a result of mobile networking, processing, storage, and other functions shift from a network to its edges (the terminals using it).

A likely vision for future networks is grouping very high capacity (several terabits per second) pipes with meshed structures. The structure will ensure high reliability and decrease the need for maintenance, particularly for costly responsive

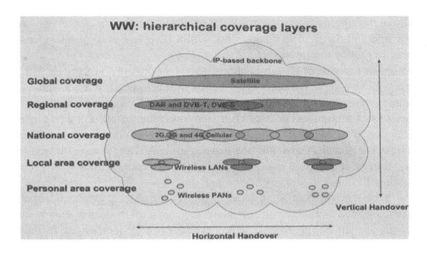

FIGURE 7.4 Femto solution.

maintenance required to maintain service quality. The network terminates with local wireless drops that will constitute a geographical hierarchy. Local wireless networks will provide coverage through very small femtocells and picocells; large WiFi cells; large planned coverage systems like FRA, LTE, 3G, and the remnants of GSM; even larger cells covering rural areas (WiMax); and coverage provided by satellites (Figure 7.4).

The wireless areas created by terminals for providing significant local coverage are now known as halo nets. A set of halos extends the terminal effect into neighboring spaces and creates a local communication fabric that flanks the terminals of the main communications infrastructures. Imagine a car creating a radial halo net extending 50 meters in every direction. The net will overlap halo nets created by other cars. As a result, an urban area can be completely covered by a wireless network in which the cars are the nodes. The network can increase its capacity; more cars will add network capacity.

Two crucial aspects of this vision are (1) ensuring seamless connectivity and services across a variety of ownership domains (in principle, each drop may be owned by a different party), and (2) vertical and horizontal roaming across different hierarchy layers rather than along cells in the same layer. Authentication and identity management are also crucial. This kind of evolution will require more transparency of the network to services. The overall communication environment will consist of millions of data and service hubs connected by very effective (fast and cheap) links.

How can we speak of "millions of data and service hubs"? The current trend is toward fewer data centers because storage and processing technologies make it theoretically possible to have a single data center for the world. However, reliability requires several replications in different locations so the end result is still several units. In the future we will see the emergence of data fragmentation storage. Every cell phone and every home media center can be seen as a data hub. The estimate of millions of data hubs is actually very conservative. How can we consider a terabyte of storage in a cell phone, 10 terabytes in a media center, and several exabytes in a network service data centre on the same level?

From an economic viewpoint, if we do the multiplication, the total storage capacity in terminals far exceeds the capacity of a network service data center (terabytes × Gterminals = 1000 exabytes). The terminals have the advantage of economics of value. Data stored in individual cell phones is of more value to the users than data stored elsewhere. Cell phone users will consider network data storage as an important back-up. Synchronization of data will take care of reliability. At the same time, asynchronous (push) synchronization from networks and service databases to terminals will make the centralized facilities invisible.

The same expansion is happening for services. Services are produced everywhere, make use of other services, data, and connectivity and are perceived locally by the users. They are bought (or may be free, possibly because of an indirect model generating revenues for the service creator to cover operational costs). Services may be "discovered" on the open Web or found in specific aggregator places. An aggregator usually adds a mark-up to the service while providing some type of assurance to end users (e.g., the Apple and Android Stores).

Assuming we have a network based on interconnected data service hubs, one of which is in our possession and another is at home, what communications paradigms are used?

Point-to-point communications (calling specific numbers) will be replaced by person-to-person or person-to-service (embedding data) communications. This will represent a departure from present communications. We will no longer call a specific termination point (telephone number). Instead, we will be connected to a particular value point (person or service). Conceptually we are always connected to that value point; communication means acting on an existing connection. The fact that the action may require setting up a path through one or more networks is irrelevant to the user, particularly if the action imposes no cost on the user. The concept of a specific number—and a strong asset for a telecom operator—will disappear.

The contextualized personal information will be stored in a "sticker" communication device. An individual or device will always be implicitly or explicitly connected with certain information. Most information may reside in a terminal, but a certain segment will relate to the site where the terminal operates or to new information generated somewhere else. Communication operates in the background, ensuring that relevant information is available immediately when needed. It will require continuous synchronization of a user's profile, location, and ongoing activities. The device will embed concepts like mash-ups of services and information and metadata and metaservice generation and require value tracking and sharing. It may require shadowing (tracking data generated by it or other terminals with which the user interacts).

The variety of devices available for communications in an environment, some belonging to a specific user, some shared by several users (e.g., a television), and some that may be "borrowed" by someone other than the owner, can be clustered to provide ambient-to-ambient communications.

Autonomic systems will help make such communications possible and even routine. A personal interaction point will "morph" into a multi-window system allowing a choice of specific windows to use for a certain communication. A user at the other end will be able to choose a way to receive the communication. Between transmission and receipt, one or more communication links that do not connect the two parties may process information from another source and utilize it within the system.

Such communications will appear spontaneous (simple) to the parties involved and more complex when executed by a communications manager. The manager can, in principle, reside anywhere. Network operators are likely to provide this communication service. Contextualized communication will become the norm in the future and represents a significant departure from the model of the communications we now use. We will explore this issue further in the final section of this chapter.

7.10 BUSINESS ECOSYSTEMS

Based on previous discussions, the future of telecommunications will consist of many loosely connected players that will create and exploit innovation. Several economists and technology experts are already wondering whether the usual relationships among participants in certain business areas can continue to be modelled on the bases of value chains.

The present consensus is that value chain modeling will be complemented by a broader view that considers business ecosystems. A vaue chain is characterized by a set of contractual obligations undertaken by the participants. Competition may cause one actor to discontinue its relationship with another and connect to a new participant that offers better quality or price for the same product or service. Value chains tend to be efficient due to the competitive value of each player in the chain. Innovation is pursued to increase efficiency. In a competitive market, over time, the value produced by efficiency at any point in the value chain tends to move to the end of the chain so that the end customer reaps the benefit.

Those sustaining the cost of innovation may see increased margins for a time, but the long-term benefit is remaining in the value chain and maintaining competitive status. Patents may protect innovations and preserve their values to their owners. This is particularly true for manufacturers but has little application to service providers. Services are easier to copy and thus circumvent patents.

An ecosystem is characterized by a loose relationships among its members who may not even know each other. In essence, an ecosystem is a set of autonomic systems in which each player manages its own game and reshapes its behaviors and interactions based on the functioning of the system or the player's specific segment.

Innovation can occur anywhere in an ecosystem and benefit any player because it increases the perception of value of the ecosystem for all players. The party that generates the innovation retains the value, usually through a direct link to the end user. However, innovation is much more tumultuous and may cause interest to shift rapidly from one player to another.

The pressure to innovate arises from end users rather than from members of an ecosystem. The crucial point for innovators remains unchanged: taking innovations to market.

The usual tactic is to "piggy back" on existing connections to the end market by exploiting advertising, distribution, and other advantages. These connections are also known as control points because of their power in bringing innovation to the end market and controlling the flow of value to users. iTunes is such a control point. Set top boxes are also examples of control points although they are parts of a value chain, not an ecosystem. That is starting to change with the appearance of "connected television" and other open applications.

Vehicle manufacturers have formed very strong value chains and tuned their effectiveness to incredible levels over the years through tactics such as just-in-time production and co-design. Their value chains led to the development of an ecosystem for production add-on components such as stereos, seat covers, and other accessories. The companies producing these add-ons simply piggy back on existing car models to offer their products without contractual obligations to car manufacturers. As these examples show, ecosystems already exist in industrial societies. The new features include the increased flexibility and openness provided by objects embedding microprocessors and software. It is now possible to add features inside an object, not just on the outside.

The computer industry is another example of the emergence of ecosystems. It exhibits a strong value chain—computer components to manufactured products to operating systems to applications. Most applications are produced by independent players who take advantage of the market created by the computer industry. Some of these applications are part of the value chain (Microsoft's Office and a similar Apple product). Others are the results of investments by independent players who choose a platform (Windows, OS X, Linux, Symbian) depending on their evaluations of their potential markets. Some of these applications utilize "plug-in" systems developed by other players and thus expand their markets further.

In these last five years we have seen the explosion of this model with hundreds of thousands of applications on both IOs and Android. Within two to five years, we will see more new platforms for supporting applications. We already see signs that applications can increase the photographic capabilities of digital cameras.

All applications and plug-ins that provide ways to refine results produced by others increase the value of an ecosystem. Sometimes, this value is so high that customers are "locked in" from moving to another ecosystem. For example, a user moving from a system based on Windows to one based on OS X would lose a number of valuable applications and find the move unacceptable.

Lock-in is a characteristic also found in bio-ecosystems. Business ecosystems and bio-ecosystems exhibit a number of similarities resulting from ground rules that apply to all complex systems involving interactions of various components. Furthermore, interactions follow various paths and produce different results based on local status of the ecosystem. "Local" is defined as the part of an ecosystem perceived by the participants of an interaction at a particular time.

The future will bring many more autonomic business ecosystems to the fore as result of the openness of objects and their flexible behaviors and interactions based on microprocessors, sensors, actuators, and software, but this is not enough. One single Roomba in a living room does not create an ecosystem. The requirement is to reach a certain number of thresholds (numbers of actors). Billions of open cell phones interacting directly and indirectly will create a huge ecosystem of enormous value—well beyond the total values of all the cell phones.

Another way of looking at the increased value produced by independent players in an ecosystem is evaluating mash-ups that can aggregate in an ecosystem (e.g., Google maps). Most mash-ups involve aggregation of information. Future mash-ups will provide both services and information. They do not cover total ecosystems since they do not represent devices like iPod ecosystems in which participants produce external speakers, pouches, and decorations. Note, however, that as objects form autonomous systems it will become progressively more difficult to distinguish the business atoms from the bits.

This observation leads us to explore the future of Internet in terms of value growth, not its physical and architectural underpinnings. Today's Internet consists of an endless store of information. In recent years, it has expanded to provide services to the point where many established actors in the service business are starting to reconsider their business strategies.

The "silver bullet" applications (e.g., televisions *inside* cell phones) sought by providers appear unlikely. Telecom operators that prospered from connectivity and progressively added other value-added services in their closed, walled markets now face a growing number of small service providers. The sheer number of small providers is sufficient to guarantee a level of innovation that exceeds the results of massive efforts by telecom operators.

Small organization are not bound to principles or plans. They simply offer their wares at low cost, leaving the broad audience of Internet surfers to try their products. The release of a "beta version" service is beyond the culture of a telecom operator. The disclaimer that the provider has no responsibility for the proper functioning of a service is unheard of in the telecom world. The Internet that introduced us to the concept of "best effort" in connectivity is now creating the same culture in the service area.

The mass market responded well to this offering. It is so wide that one can find an interested party for basically any proposition it receives. Only a few niches may be interested in acquiring and using a service (application at present) but these niches span the planet, and they are sufficient to generate returns on the (usually small) investments of those who created the services.

Although connectivity is key in enabling service access and use, the value from the user view shifts from connectivity to service (and to terminal when the terminal is seen as the service enabler as is the case for the iPhone and Galaxy). Note that the trend toward embedding information into services is a further drive in this direction.

Now we type to enter a browser to access a site to get a weather forecast. It is a very different task to click on a sun symbol on an iPhone and obtain forecasts for places you care about. No connectivity is involved in this scenario. A user may not know whether the information results from a click to activate a network connection or was pushed to your device as you recharged it. A user may not care what kind of connectivity was used to bring the information to the terminal (one of the thousands of WiFi networks to which the device can automatically and seamlessly connect or a cellular network). Communication is no longer an isolated task.

The attempts of operators to charge for connectivity on a bit-by-bit basis is losing ground and competition is already arising. Flat rates imposed on mobile data have become customary for fixed line access. As a result, operators will no longer see revenue increases from connectivity or from future "killer" applications. Market penetration in most Western markets has reached a point where further market growth is unlikely and progressive squeezing of margins is almost certain.

7.11   INTERNET WITH THINGS

Actually, the number of humans in those markets cannot grow much further, but what about objects? We have seen how technology makes it possible to embed intelligence in objects and enable communication among objects and between objects and humans. The problem with objects from a business view is that they do not have check books or bank accounts—they cannot pay for communications or services.

As a result, the "Internet of Things" (machines connected to machines via the Internet) is already a state-of-the-art system and the number of connections is growing. The "Internet with Things" (Figure 7.5) is the big market in terms of value generated.

We can say the Internet contains everything (someone already called it the "Internet of Everything" (IoE). Humans with wallets can access and find value in information, services, and objects in a continuum. It is becoming more difficult (and irrelevant) to separate one from the other. Does this evolution have the capacity to increase the overall value and free resources that can be monetized by operators and other actors?

Any object can become a node for delivering services and information. A cell phone can act as a bridge between the atoms of an object and the bits of related services and information. The key enabler is the unique identification of the object. The identification can be used as an address to procure services and information, created and managed by various and independent players in the true spirit of business ecosystems. This simple mechanism will change our view of the world by welding the virtual and real worlds together. A cell phone may become a magnifying lens for placing Web information onto any object.

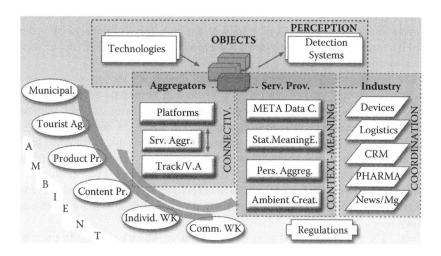

FIGURE 7.5   Internet with things.

The connectivity and flexibility of objects will allow producers to keep in touch with their products and update or upgrade them as new features become available. If an object is "open" (most will be) the features increased by third parties may greatly improve the versatility of the object and the value perceived by the user. Customer care will take a completely new twist as it shifts from remotely fixing problems (and that will become much more feasible in the future) to a way to increase the user experience. In many instances, customer care will be provided by third parties. It will not be outsourced by producers. Customer care will be handled independently of producers by organizations that treat the service as a business opportunity. Objects will be able to interact and thus create environments that are aware and improve the experiences of the people living in the environments.

Telecom operators will be able to act as intermediaries for these completely new businesses and may reap fresh revenues from them. Basically, telecom earn their revenues from intermediation between people via connectivity. Now that they are running out of human customers, they will conduct intermediation with objects and we know there are so many more objects than people. The objects will not directly support business by becoming subscribers. Telecom revenues will result from the exploitation of the mash-ups of the virtual world onto the physical one.

Production, delivery, and post-sales chains will become much more efficient, less costly, and thus will free money that may at least in part go to operators. Clearly, this new market will open to all and competition will be fierce. Some of the assets of telecom operators may give them some competitive advantage in this new game, and it will be impossible to succeed by playing the game with the old rules that succeeded in the past. The optical network and its wireless drops will serve as the enabling infrastructure to sustain the traffic increase and the flat rate tariff without which the market is unlikely to take off.

7.12 COMMUNICATIONS IN 2020 (OR SOONER)

We have reviewed a few technological areas, discussed their evolution, and seen the evolution of the business framework with the emergence of the ecosystem paradigm. An interesting question is how all these advancements will affect our way of communicating.

Most terminals will be wireless, usually cell phone-type devices with 3-inch (or more) screens with good resolution. Top-of-the-line terminals will probably use OLED technology that features a very bright screen visible in sunlight, with very small pixels to pack high resolution onto a small surface. The backs of some models may have black-and-white screens based on electronic ink technology so that more ambient light will improve image quality. Some devices will provide flexible screens that can expand to create larger visualization surfaces. These terminals are better described as virtual lenses that will reveal items that may be hidden at first sight, somewhat like the magnifying lenses that were fashionable in the seventeenth century.

Terminals will be equipped with sensors, some of which will identify users who handle them. When a user picks up a screen, it will identify the context (usually location) that may be represented as a background image. The user's face will appear at the center of the screen. If he or she hands the device to another person, that person's face will be on the screen.

This simple act of handing the device to another person will require access to the identity and associated data of that person via a network. The use of cloud technology is a step in this direction. Alternatively, identity may be determined by some local personal identification system (chip embedded in the body as in the Intel WISP or a local dongle that acts only when in close proximity to the authorized person. As an example, if I pick up a cell phone, it becomes mine if the owner authorizes it; otherwise the phone will remain inert and useless.

When I use the phone I see a number of icons, some automatically brought forward from earlier cell phones. The transition to a new phone model requires no work; it happens seamlessly on any cell phone I own or am authorized to use. This feature is a reality now with iCloud. These icons are the new communication features of this decade. Plenty of examples are already visible in today's smart phones, although they will look very primitive within a few years.

The missing feature is the speaker. New devices are based on the concept of visual communication by looking at a screen, Visual communication is impossible if a phone must be placed over a user's ear. Sound from a new device will reach your ear via a classic Bluetooth ear plug or a more sophisticated sound beaming device. Sound beaming systems are available but they require several speakers to focus the sound in a small areas. Progress in processing capacity and smart materials will eventually minimize the sizes of speakers in phone shells.

Keypads will also be eliminated. The point is that all communication (with very few exceptions) will be contextualized, for example, like entering a room and interacting with a person or object. Setting up icons to identify parties you call may seem to require an impossibly high number of icons and recognizing them may be challenging. What if you want to call a new person for whom your phone has no icon?

The present number dialing system requires input of 9 to 13 digits and lets us reach some billion terminals. On a 3.5-inch screen like the one on an iPhone, a user can place twenty icons and have the same number of choices with only seven clicks. In terms of pure clicks, we can replace numbers by icons but this does not solve the issue of identifying the right icon.

The problem can be solved by a tree structure in which every click basically restricts the context. Suppose I want to call John, a friend and tennis partner. I can click on the *community* icon, then on *tennis*. John's face will appear in an icon. I might also start by clicking *home*, *agenda*, and *Wednesday last week*. John's face will again appear on an icon because we had dinner together last Wednesday. These examples demonstrate how our brains remember—through association. The best interface we can devise is based on association (and that is not the one we use when dialing numbers).

This issue of identifying something or someone I want to call is interesting. It requires more than an interface. It also involves data structuring, architecture, and possibly several players in an ecosystem approach. Suppose you want to call a restaurant you visited last month with a business associate but do not remember its name. Simply click on your face and on *history*. Drag the arrow of time to the past month and you will see a collection of memories stored in the phone including the dinner at the restaurant. You did not write down the name of the restaurant because your associate took you there, but your cell phone has the location tracked and stored. You simply have to click on the restaurant to view it.

The location information will be sent to a directory provider (that actually will do more than provide directory information). The directory will push onto your screen a new context containing the restaurant icon that would allow you to see inside the restaurant, browse the menu, call the restaurant or make a reservation, and arrange dinner dates with several friends. The transportation icon will let you know how to get there. You will even be able to view comments about the restaurant left by other patrons.

It is easy to see that an interface based on context really represents an interface to human thinking. Its implementation requires the generation of various types of data owned by many parties located in several places. A service provider that knows how I think and has my profile will reshape the data. Some of the transformation may take place inside my terminal that can store 1 terabyte of data and is very knowledgeable about my activities. To answer my questions to the outside world, the terminal may even decide to mask my identity for privacy reasons. Personalized and contextual communications will become realities in this decade.

Let's look at some icons on a typical screen. One may be a shopping cart. Click on it and you'll see a number of new icons, one of which is a to-do list that shares information with your family or business associates, depending on the context you are using. If you are using the family context and your spouse buys bread, the bread item disappears from the list. If you add an item to the list, the item is added to the lists of other family members. No action is required to revise the list. As a family member uses a phone to pay for bread at the supermarket the item is deleted from all the connected list. We already see this phenomenon in iCloud's seamless synchronization of calendars across terminals.

The shopping list may be shared with your car. As you drive, the car neutralizes your identity, makes inquiries of stores along the way, and will prompt you to the availability of items in nearby stores. You may elect to let the sharing occur on a broader environment and accept ads for certain items so you can review a variety of offers. You would not use this capability to buy bread but it would certainly be of interest if you wanted to buy a digital camera or rent a property.

The control of information sharing is going to be crucial both for its acceptance and for monitoring advertising value. One or more third parties may be involved in control. So many potential actors are involved and the margins are so razor thin that most processing should be handled automatically by the communications fabric that intermediates among the players.

Another icon shows your previous shopping sprees and those of family members who want to share the information. How is this information captured? Clearly any purchase made with a credit card is potentially traceable. However, obtaining it may be close to impossible because of the number of databases involved and the variety of security systems and ownership domains. Life (at least in this respect) would be simpler if a cell phone could be used for any transaction. Note that this icon does not reveal details of transactions; it simply keeps track of purchase. Perhaps a future operator will develop a complete tracer system that customers will be willing to use to keep track of their lives.

Another icon will cover shops in the area. One click will display several icons of shops in the area that pushed themselves onto your screen. Can they do that? A platform is required to manage the information and the formatting of applications appropriate to the visualizing terminal must be in place. Some type of agreement from the cell phone user will be required to permit the icons to appear on his or her screen and certainly the user should not bear the cost of accessing these advertising icons.

This possibility of entering in the menu of cell phones is a very interesting one for retailers who will have a way to get in touch with potential customers roaming in their area.

Consider another first-level icon that represents my home. A click on the call-home can activate a connection to the home but probably it is more interesting to stretch the icon to fill the screen. This allows the home to become my context and I will have a choice of icons for service. The icons will allow me to interact with people and appliances that are part of the home environment at that moment.

I like to close this chapter with a description of one more icon that takes the shape of an eye because its objective is to transform a cell phone into a lens to view information and services layered on an object. Such devices are already in use in supermarkets in Japan and on billboards in France. A tag on the object can be read by a cell phone camera that then retrieves the unique identity of the object. It is already becoming common to see QR codes for cell phone applications along with bar codes on products. The identity tag can be used by interested parties to retrieve relevant services and information. Every object will have the potential to become a point for accessing and distributing information and services and join the Internet with Things.

The cell phone is likely to be the main intermediary in communications activities and the point of aggregation of personal information for customizing services. However, we will have more opportunities to communicate beyond those offered by cell phones. An individual's environment (home, hotel room, office) will be his or her communication gateway to the world. Walls will display information. Sensors will customize the environment to an individual's taste and cameras will be able to project images to far distant places.

Any object can be overlaid by information in a manner similar to the way we affix a label to a box or underline text with a crayon. Objects transported by carriers such as FedEx or sent in bits like digital photos will contain information a receiver can hear. Communications will utilize innovative wired and wireless systems that in many case will replace the classical communications infrastructure.

Among all these possibilities, I am certain about one: communication will be the invisible fabric connecting us and the world whenever and wherever we are in a completely seamless way. Communication will be so transparent, inexpensive, and effortless that we will seldom think about it.

8 OCTOPUS: On-demand Communication Topology Updating Strategy for Mobile Sensor Networks

Hicham Hatime, Kamesh Namuduri, and John Watkins

CONTENTS

8.1 INTRODUCTION

Mobile sensor networks are characterized by frequent topology updates and self-configurations. Networks are created spontaneously whenever nodes are within transmission range of each other. The arrivals and departures of nodes in a network are ongoing dynamic processes because of their high mobility. This dynamic nature causes the nodes in mobile sensor networks to lose connectivity. Strategies are required to keep a network connected and adaptable to frequent changes.

Although suitable for many applications, the unpredictable overhead generated by frequent topological changes can be cumbersome. To handle this problem, some strategies focus on node power control mechanisms [1–4] while others emphasize node hierarchy [5–8]. Power control schemes tune power at each node to ensure the closest neighbor connectivity and thus the overall network connectivity.

The hierarchical-based approach, also known as clustering, consists of dividing nodes into two subsets: (1) a set of cluster heads and (2) a set of nodes associated with cluster heads. Connections of cluster heads are established through selected gateway nodes. Regardless of the approach, the challenge is to keep a network stable as long as possible before reorganization becomes necessary.

In this chapter, we propose a periodic and on-demand topology reconfiguration framework for a mobile sensor network that results in enhanced connectivity and performance. The reconfiguration occurs on demand when nodes leave or join the network and periodic when employed as an alternative approach to topology control and maintenance through scheduled reorganization.

The proposed approach consists of four stages. The first stage utilizes a combination of adaptive resonance theory (ART) and Maxnet to cluster nodes. Maxnet's weight update equation helps find adjacent nodes in the network. The ART approach contributes to the formation of new clusters when adjacency conditions are not satisfied.

The second stage addresses the election of a cluster head for each cluster to facilitate intracluster communication. The selection is performed using an integer linear programming formulation on the adjacency or neighborhood matrix.

The third stage deals with the establishment of intercluster communication. Gateway nodes are selected based on a willingness function that combines node degree, distance to cluster center, available power, and transmission energy. The fourth stage consists of establishing optimal connections between clusters and is also performed using an integer linear programming formulation.

This chapter is organized into seven sections. Section 8.2 describes related work on this topic. Section 8.3 presents the mathematical concepts used in the proposed model. Section 8.4 outlines the proposed on-demand communication topology update strategy. It also describes the neural network approach for clustering nodes based on range and proximity and the selection of cluster heads using the concept of minimal dominant sets. Other topics covered include the willingness function for selecting the gateway nodes used for communications among clusters and an approach for cluster interconnection. Section 8.5 analyzes the complexity of the method. In Section 8.6, simulation results are discussed. The final section summarizes conclusions and proposes future work.

8.2 RELATED WORK

Different heuristics to create clusters, elect cluster heads, and select gateway nodes have been proposed in the literature. Some systems use a node identifier to elect cluster heads. This approach, also known as lowest ID heuristics or identifier-based clustering, was introduced by Baker and Ephremides [10]. It assigns a unique ID to each node and elects the one with the lowest ID as a cluster head. A cluster is then created by connecting all nodes with higher IDs to the cluster head. A node is selected as a gateway only if it lies within the transmission range between two cluster heads.

Clustering based on node degree, max degree, or connectivity, is another common approach introduced by Gerla and Tsai [11]. A node with higher degree is more likely to be elected a cluster head. Using any protocol for neighbor discovery, a node with a maximum number of neighbors is elected. If a tie occurs, the node with lowest ID is selected. The neighbors are then connected to the closest cluster head to form a cluster. Only one cluster head per cluster is allowed and cluster heads are connected to each other to act as gateways.

Another approach is to assign weight to a node based on its readiness and disposition to become a cluster head [12]. A node is elected a cluster head if its weight is higher than the weights of its neighbors. A number of methods can assign weights based on node energy, position, degree, speed, direction, and other factors. Others use a probabilistic approach.

Tan et al. proposed a priority-based adaptive topology management approach that uses a heuristic weight function based on distances to neighbors, current communication round, energy level, and node speed [13]. The node with highest degree becomes the cluster head.

Heizelman et al. proposed a low energy adaptive clustering hierarchy in which each node uses a probability value to elect itself a cluster head [14]. This approach is based on the number of cluster heads and their randomized rotation. Most of the systems cited in the literature are power- or hierarchical mechanism-based. One condition for a connectivity and topology strategy is to encompass and balance the parameters of the two mechanisms. Neglecting or favoring one set over the other generally leads to a weak strategy.

Cluster formation, cluster head election, and gateway selection have been covered extensively in the literature. However, an on-demand topology reconfiguration framework suitable for a mobile sensor communication network that combines parameters from hierarchical and power mechanisms is not included in the literature.

8.3   MATHEMATICAL PRELIMINARIES

In this section, we present the assumptions and the mathematical concepts used in the proposed model. First, nodes must be confined to a geographic area. Nodes outside the defined perimeter belong to a different area. Communication between areas is not considered in this chapter. The assumption of a geographic area allows nodes to find each other via broadcast messages or through neighbor discovery—not a trivial task.

Ideally, an omnidirectional antenna with variable transmission range to cover the confined area would be appropriate. This is necessary since nodes have to adjust their transmission ranges for intercluster communication. An alternative solution is a directional antenna that is likely to detect more neighbors because of longer transmission range and also requires processing antenna direction. For distributed processing, nodes are assumed to determine their respective locations and include the data in their advertisements and communications along with their transmission ranges. Nodes should also share knowledge about other nodes' positions and transmission ranges.

The second assumption requires that nodes communicate through their elected cluster head. Communication among peers is not allowed. Instead, communication between cluster heads is performed via a designated gateway. A back-up gateway, if available, will act as a normal node to prevent loops. In idle state, nodes have limited transmission ranges. The proposed approach uses this limitation during cluster formation and assumes that the nodes will adjust their transmission ranges later.

We represent the network of nodes as an undirected graph $G = (V, E)$ where V is the set of vertices and E is set of edges. The neighborhood $N(v)$ of a vertex v consists of a set of vertices adjacent to v, that is, $N(v) = \{u \in V: uv \in E\}$. Thus, the adjacency matrix A and neighborhood matrix N can be represented as:

$$A = [a_{u,v}] = \begin{cases} 1 \ if \ uv \in E \\ 0 \ otherwise \end{cases} \quad\quad (8.1)$$

$$N = A + I_n$$

where I_n is the identity matrix.

We also use the concept of domination in graph theory. A node u dominates another node v, if u and v are linked. Each of the several ways to define a dominant set in a graph illustrates a different aspect of dominance [18,19]. The definitions we followed are described in Reference [15] and will be used in Section 8.4.2. Throughout this chapter, CH denotes a cluster head, G a gateway node, and N a normal node.

8.4 PROPOSED MODEL

This section describes the steps involved in building a communication topology for a randomly placed group of nodes. In the first step, nodes within range of each other are grouped into clusters. In the second step, cluster heads are selected to facilitate intracluster communication. In the third step, gateway nodes are selected to facilitate intercluster communication. In the fourth step, intercluster links are established.

8.4.1 CLUSTERING

Clustering is the process of partitioning a set of nodes into subsets that share a common characteristic. The most common criterion for grouping nodes is proximity. The two methods for determining how nodes are organized are supervised and unsupervised. We focus only on unsupervised learning and particularly on competitive neural networks (NNs). We combine the concepts of two NN models (Maxnet and ART) to develop a clustering strategy.

Maxnet [9] is an NN that does not require training, serves as a classifier, and is used to identify the winning node with the highest weight. In our implementation, we wanted to find nodes that satisfied the constraint of being within range. Our approach starts with assigning initial weights to links based on the distance between nodes.

It uses an adjacency function $f(x)$, as an indicator of an adjacent node; a weight update function to pick the "winning" node in the competition, and a stopping condition to halt the weight update [9]. The adjacency function and the weight update function were also used to evaluate the adjacency A and neighborhood N matrices. The adjacency function is described by:

$$f(x) = \begin{cases} x & \text{if } x > 0 \\ 0 & \text{otherwise.} \end{cases} \tag{8.2}$$

The weight update function is described by

$$w_{ij}(new) = f\left(w_{ij}(old) - \beta \sum_{k \neq j} w_{ik}(old) \right) \tag{8.3}$$

where $0 < \beta < \dfrac{T_i}{2}$, w_{ij} is the distance between node i and j, and T_i is the transmission range. The iterative process of the weight update given in Equation (8.3) reduces the distance between any two nodes (say, i and j) by a small value $\beta \sum_{k \neq j} w_{ik}(old)$ and tests whether the weight reduces to zero. Note that $f(x)$ also serves as a stopping condition. The weight update function requires that the quantity subtracted satisfies the inequality $\sum \left(\beta \sum_{k \neq j} w_{ik}(old) \right) \leq T_i + T_k$. A value of zero in any row of the weight matrix (w_{ij}) indicates that a within-range adjacent node has been identified.

The second concept borrowed from ART [9] allows the user to control the degree of similarity of patterns placed in the same cluster and in turn control the number of clusters. New inputs are first tested against the existing clusters. If no cluster fits the new input, a new cluster is created. Inputs with similar patterns are grouped in the same cluster. The creation of new clusters relies on a vigilance condition defined as follows.

$$Dist(i,j) \leq T_i + T_j. \tag{8.4}$$

The algorithm creates a cluster for each pattern that does not satisfy the vigilance condition. Note that the right side of the vigilance condition can be controlled by the user.

8.4.2 Computing Minimum Dominant Set

After evaluating the adjacency matrix and determining the clusters in step 1, the next step is to elect the cluster head set S using the concept of minimum dominant set (MDS). Computing MDS has been an active area of research in many fields. Among the earliest formulations of the problem was the placement of the five queens in a chessboard game. The queens must be in positions to occupy or attack any square.

Many heuristics, mostly in graph theory and ad hoc networks, have been proposed to solve the problem of defining the MDS. Whether centralized [21,22] or distributed [23,24], the proposed algorithms try to reduce the problem complexity (which is NP-hard). In our approach, we use a technique that is pseudo-polynomial since it uses integer linear programming (ILP) which is known to be NP-complete [25]. In the proposed approach, MDS is computed using a characteristic function:

$$f: V(G) \rightarrow \{0,1\} \tag{8.5}$$

which satisfies

$$f(v) = \begin{cases} 1 & \text{if } v \in S \\ 0 & \text{otherwise} \end{cases}. \tag{8.6}$$

The characteristic function states that v is a cluster head if it belongs to S. This function will be used to assign values to the variables in the linear programming formulation.

There are two ways of defining the MDS [15,16], both of which yield the same results during simulation. The first approach uses the adjacency matrix in which a vertex v is considered to dominate the vertices in its neighborhood but not itself. It is equivalent to finding the vertices that have maximum neighbors and cover the whole graph:

$$AX \geq \overline{1}_n, \tag{8.7}$$

where A is the adjacency matrix, $X = (x_1, x_2, \ldots, x_n)$, is the column of variables to be evaluated, $\overline{1}_n$ denotes the column n-vector of all 1s, and n is the number of nodes. The integer programming formulation for the minimum cardinality domination set $\gamma(G)$ is described as:

$$\gamma(G) = \min \sum_{i=1}^{n} x_i \tag{8.8}$$

subject to $AX \geq \overline{1}_n$ with $x_i \in \{0, 1\}$, which translates to minimizing the number of nodes that have the highest numbers of adjacent nodes.

The second formulation which uses the neighborhood matrix, assumes that each vertex can be dominated only once and seeks to achieve as much domination as possible. Efficient domination happens when the maximum number of dominant vertices can be found. The integer programming formulation for computing the efficient dominating set $F(G)$ is given by:

$$F(G) = \max \sum_{i=1}^{n} (i + \deg(v_i))x_i \tag{8.9}$$

subject to $NX \leq \overline{1}_n$ with $x_i \in \{0, 1\}$, where N is the neighborhood matrix and $\deg(v_i)$ is degree of node v_i. This formulation translates to maximizing the number of neighbors of v_i.

At this stage, computing MDS by either method yields a set of all possible dominating nodes for the graph. As a result, some clusters have more than one dominant node. Selecting a cluster head within one cluster requires the removal of the other candidates. The process of eliminating dominant nodes within one cluster is based on node degree. The highest degree node is selected as the clusterhead. If a tie occurs, the node with the lowest ID is selected.

8.4.3 Selecting Gateways

In the previous sections, we discussed ways to create clusters and elect cluster heads within each cluster. In this section, we discuss gateway selection. The purpose of gateway selection is to establish and facilitate communication between clusters. The mechanism that drives the selection of a gateway among multiple nodes involves two steps: selection and decision.

We introduce a new "willingness function" to the selection process. The decision process for electing a gateway relies on the node with the highest willingness value. The pool of nodes is limited to those within the same cluster excluding the cluster head. The proposed willingness function incorporates four factors described below. The proposal was inspired by the technique [17,20] for electing a cluster head. We adopted it for gateway selection by incorporating the following parameters.

8.4.3.1 Distance to Centroid

This parameter denotes the distance of a node from the cluster's centroid which is defined as follows:

$$C\left(x_{Centroid}, y_{Centroid}\right) = \frac{1}{n}\sum_{i=1}^{n}\left(Rx_i, Ry_i\right) \tag{8.10}$$

where n is the number of nodes in the cluster. Rx_i and Ry_i are the x and y coordinates of node i. The distance from a node j to this centroid is described by:

$$DistToCent_j = dist(j, C) \tag{8.11}$$

where j is any node under consideration for gateway selection. A node closer to the centroid is a lesser candidate to act as a gateway and a peripheral node is more likely to be selected.

8.4.3.2 Transmission Energy

This parameter measures the transmission energy required by each node to reach all other nodes in the same cluster:

$$TrE_j = \exp\left[T_j\sum_{i=1}^{n}d_{ij}^2 \middle/ \left(1 + \sum_{i,j=1}^{n}d_{ij}^2\right)\right] - 1 \tag{8.12}$$

where T_j is the transmission range and d_{ij} is the distance between node i and node j. A node farther from all other nodes requires more transmission energy to communicate with the other nodes. Unlike the *DistToCent* parameter, *TrE* is directly linked to the required battery power.

8.4.3.3   Node Degree

This parameter reflects a node's neighbors within its transmission range.

$$Deg_j = \sum_{i \neq j}(dist(i, j) < T_j), \tag{8.13}$$

where T_j denotes the transmission range of node j.

8.4.3.4   Available Energy

This parameter is a gauge for a node's available energy. Three types of energy costs are associated with this parameter. As time progresses, a system incurs a cost for sitting idle and listening for traffic (*IdleE*), a cost for movement (*MvE*), and a cost for transmitting (*TrE*). *MvE* and *IdleE* are mutually exclusive.

$$AvailE_j = AvailE_j - IdleE_j - TrE_j - MvE_j \tag{8.14}$$

$$(MvE > TrE > IdleE).$$

The willingness function (WF) is a combination of these parameters and is defined as follows:

$$WF = 1 - \alpha * DistToCen - \beta * TrE - \lambda * Deg - \gamma * AvailE. \tag{8.15}$$

The values of α, β, λ and γ are user-defined parameters and total 1. As an example, the values assigned for the initial topology creation are $\alpha = 0.4$, $\beta = 0.1$, $\lambda = 0.5$, and $\gamma = 0$ since we assume that all nodes have equal energy at the start. For a given cluster, the willingness function is evaluated for all nodes within it. The node, excluding the cluster head (D3), with the highest willingness value is selected as the gateway for the cluster.

8.4.4   CLUSTER HEAD AND EXTERNAL GATEWAY LINK

Cluster heads and gateways are the fundamental components of a mobile sensor communication network. All communications pass through them before reaching the rest of the nodes. After the cluster heads are determined, the challenge is to find a link that connects them under the constraint that the communications between cluster heads occur via gateways. However, due to the randomness of node placement, this constraint is not always guaranteed to be satisfied.

The purpose of establishing a cluster head and an external gateway link is to connect a gateway with an external cluster head (ECH) for a complete topology. Several scenarios need to be considered. Generally, a cluster head is not a candidate to be elected as a gateway. However, there may be cases in which a cluster is formed by a single node that acts as a cluster head and gateway at the same time. To solve this problem, two separate approaches are proposed. The first tackles the normal case, and the second approach deals with the exceptional case.

The first approach, in which one gateway is linked to one external cluster head, can be compared to a perfect matching graph assignment with equal set cardinality that may be solved by this simple expression:

$$\min \sum_{i=1}^{n} \sum_{j=1}^{m} c_{i,j} x_{i,j}$$

$$subject\ to\ \sum_{i=1}^{n} x_{i,j} = 1$$

$$\sum_{j=1}^{m} x_{i,j} = 1$$

$$with\ x_{i,j} = \{0,1\}$$

(8.16)

where $x_{i,j}$ is the variable to be evaluated and $c_{i,j}$ is an element of the distance matrix between all gateways and external cluster heads belonging to multinode clusters.

The second approach requires that a gateway may connect to more than one external cluster head belonging to a single-node cluster and that each external cluster head is linked to one external gateway. The formulation of this problem can be represented as a minimum weight b-perfect matching problem and solved using the following integer programming expression.

$$\min \sum_{i=1}^{n} \sum_{j=1}^{m} c_{i,j} x_{i,j}$$

$$subject\ to\ \sum_{i=1}^{n} x_{i,j} \geq b_i$$

$$\sum_{j=1}^{m} x_{i,j} = 1$$

$$with\ x_{i,j} = \{0,1\}$$

(8.17)

where $c_{i,j}$ is an element of the distance matrix between all gateways and external cluster heads belonging to single-node clusters and b_i is the minimum number of external links allowed per gateway ($b_i = 1$ in our case).

8.5 COMPLEXITY

Analyzing the efficiency of an algorithm is of primary concern to any user. The goal is to know how many resources (time, storage) are required for the technique to execute. The objective of runtime analysis is to estimate the increase in time when the number of inputs increases and determine an upper bound or limit for execution time. Analyzing a technique necessitates analyzing each component.

The reconfiguration method has three parts: clustering, gateway selection, and intercluster link establishment. Clustering involves three nested loops $\theta(n^3)$ and solving an ILP problem. Gateway selection is accomplished by two nested loops $\theta(n^2)$.

Intercluster link establishment is computed using two ILP formulations. ILP uses a branch- and bound-based search technique [29,30] that in turn uses a tree structure to solve the problem. It partitions the feasible region into subsequent smaller subsets and then calculates bounds for each subset. These bounds are then used to discard some subsets and to update the current solution to a better one.

The iterative process stops when no feasible solution or better solution can be produced. In general for $n \geq 3$, the optimal solution will result in $(n - 1)/2$ decision variables x_i equal to 1 and $(n + 1)/2$ decision variables x_i equal to 0, suggesting that ILP is at least $\theta(2^{(n+1)/2})$ exponential. The approach has an exponential order and is suitable for applications that require small numbers of nodes.

8.6 EXPERIMENTS AND RESULTS

The proposed on-demand topology update strategy is simulated in MATLAB®. In this section, the network topology formation is explained step by step. To test our approach, a series of experiments were conducted with different sets of nodes randomly placed in a geographic area. Several distributions were tested to adequately model the node positions.

Randomly generated positions were evaluated using 32 distributions. Only four appeared to model the data with a confidence level above 95%. Table 8.1 shows the goodness-of-fit values for a Kolmogorov-Smirnov test. A p-value (test significance) above 0.05 indicates that the distribution can model the data. A higher p-value suggests a better distribution.

For a set of random positions, Figure 8.1 shows the results of cluster formation, cluster head election, and gateways selection for eight nodes with equal transmission ranges. Four clusters were identified. Cluster 1 comprises nodes CH7, N1, N4, and G6. Cluster 2 is formed by CH5 and G3. Finally, CH8 and CH2 form clusters on their own.

TABLE 8.1

Goodness-of-Fit Values for Kolmogorov-Smirnov Test

Distribution	Smallest Extreme	Log Logistic[a]	Triangular	Normal
p-value	0.91	0.69	0.67	0.38

[a] Three parameters.

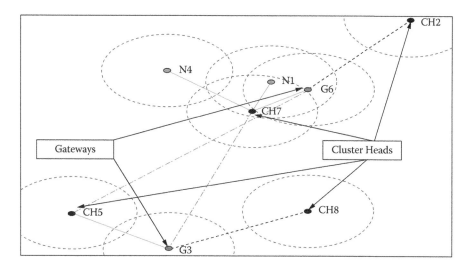

FIGURE 8.1   Cluster head and gateway selections with four clusters.

A node labeled CH is the elected cluster head for a cluster. A node labeled G is the gateway; N is a normal node. In the case of single-node clusters, the cluster head also plays the gateway role. In the figure, CH7 and CH5 are cluster heads that use gateways G6 and G3, respectively.

Isolated clusters must establish links among themselves. As a result of applying Equation (8.17), a link is established between a single-node cluster and a gateway. CH2 is assigned to G6 and CH8 is connected to G3. Figure 8.1 illustrates the establishment of the links between the single-node clusters and gateways.

The final step is to establish a link between the multinode clusters via their respective gateways. Using Equation (8.16), CH5 is linked to G6 and CH7 is connected to G3. Figure 8.1 illustrates the establishment of the link between multinode clusters. However, unlike the connections shown in Figure 8.1, not all gateways will be connected to all cluster-heads and vice versa. The implementation is such that an optimal connection is established among the set of gateways and cluster heads. A subsequent Figure (8.4a) shows a fault-tolerant connection between three cluster heads and three gateways. No link between CH3 and G2 or CH4 and G5 is established.

Since the nodes are positioned randomly, situations occur in which all nodes form clusters on their own. In those cases, a single link is established between the cluster heads. The major concern with these topologies is the cost of a link failure. Any loss of connection will automatically require a topology reconfiguration, unlike a link failure between a gateway and a multinode cluster. In general, many configurations show that no topology update is required if a multinode cluster loses one of its intercluster links.

Certain factors may affect the overall structure of a communication network: the number of nodes and the transmission range. Figure 8.2a shows that increasing the number of nodes is followed by a decrease of the number of clusters. In general, saturating an area with a higher number of nodes causes the number of clusters to

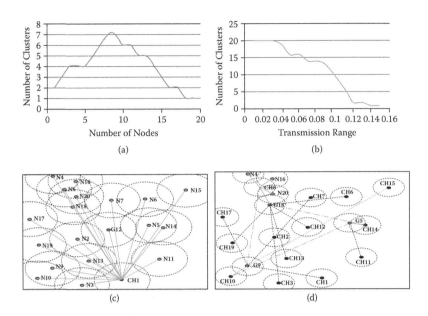

FIGURE 8.2 Effect of number of nodes and transmission range on number of clusters.

decrease. This is illustrated in Figure 8.2c, where 20 nodes with equal transmission ranges form a single cluster with CH1 as a cluster head and G12 as a gateway. This configuration negatively impacts the power and workload of the cluster head. Assuming that the probability of a link or node failure is minimal, the call for a topology reconfiguration will result mainly from low battery power to a cluster head.

Figure 8.2b shows that reducing the transmission range increases the number of clusters. As illustrated in Figure 8.2d, the same 20 nodes with lower transmission range resulted in 14 clusters and 3 gateways (G18, G5, and G9). Compared to Figure 8.2c, this configuration requires more processing and computing time to establish links of all the clusters.

Another experiment involved the distribution of intracluster and intercluster links established between nodes. The purpose was to measure the quality of the clustering technique and the goal was to obtain dense intracluster links and sparse intercluster links. Figure 8.3a shows that the ratio of intracluster to intercluster links increases with an increase in transmission range. Figure 8.3b demonstrates a similar result for an increase of the number of nodes in a network.

Dynamic transmission range is another factor that affects the structure and the role of a node in a network. Figures 8.4a and b show changes in node positions with different transmission ranges. The result is a new communication network composed of four clusters in which some nodes assume different roles. Nodes 1 and 7 were, respectively, a gateway and a cluster head in Figure 8.4a and serve as normal and gateway nodes, respectively, in Figure 8.4b. Testing the capabilities of the approach with random positions and random transmission ranges was both tedious and challenging. It demonstrates the need to keep the transmission range parameter constant for all nodes.

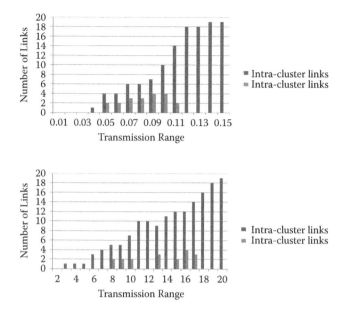

FIGURE 8.3 Transmission range and number of nodes versus density of intracluster and intercluster links.

To compare the performance of our technique (OCTOPUS) in terms of clustering capabilities, we conducted two scenario experiments similar to the one proposed by Chen et al. [31]. Three clustering techniques were selected: hierarchical, subtractive, and quality threshold.

Hierarchical clustering [32] relies on an agglomerative algorithm to find clusters. It considers each node as a single cluster and combines them to find a final number of clusters. Subtractive clustering [32] computes the likelihood of a node as a cluster head. The nodes in its vicinity are considered neighbors. Quality threshold clustering [32] requires definition of diameter for a cluster such as the maximal distance between two nodes. It relies on adding a candidate node that minimizes an increase of cluster diameter. None of these techniques requires defining the number of clusters in advance.

The experiments involved increasing the number of randomly placed nodes in a small area and applying the four clustering techniques. The number of clusters and the average cluster size were the two parameters selected to describe clustering capabilities. OCTOPUS exhibited better clustering behavior. Figure 8.5 indicates that OCTOPUS consistently maintains a lower number of clusters than the other techniques. Its curve indicates that the number of clusters tends to decrease when the number of nodes increases as noted previously. In Figure 8.6, OCTOPUS tends to merge small clusters and favors big cluster sizes; the other techniques create smaller clusters.

Although OCTOPUS outperformed the quality threshold technique, their curves in Figure 8.5 suggest that they exhibit similar behaviors. Based also on appropriate

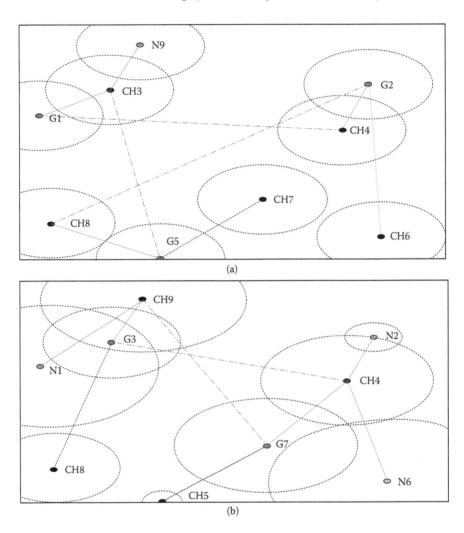

FIGURE 8.4   Effect of transmission range on role of node.

diameter value, the techniques may be similar. Figure 8.7 illustrates the behavior of the quality threshold technique when implemented with different diameter values. As the diameter increases, the quality threshold approaches OCTOPUS behavior. Thus, OCTOPUS clustering could be assimilated into quality threshold clustering with a dynamic diameter adjustment.

A comparative experiment was conducted to evaluate the overall behavior of our deterministic approach against a probabilistic approach. Barolli et al. introduced a probabilistic cluster head election technique based on the available power, node degree, and distance to the cluster center [33]. Zhang et al. proposed a gateway selection based on a probability p [34]. Combining the two approaches resulted in a complete probabilistic selection technique for gateways and cluster heads. Scores

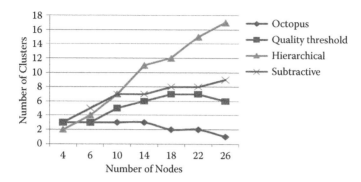

FIGURE 8.5   Numbers of clusters generated by different techniques.

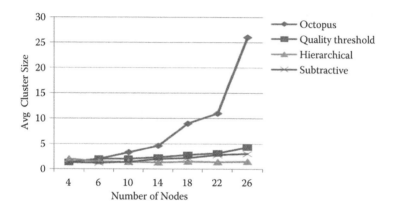

FIGURE 8.6   Average cluster sizes generated by different techniques.

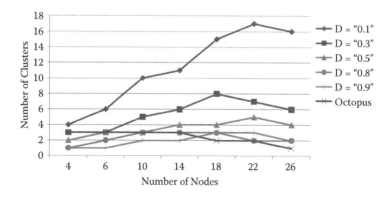

FIGURE 8.7   Quality threshold behaviors with different diameter values.

of random topologies of 10 nodes were generated and 2 parameters were evaluated using OCTOPUS and the probabilistic approach.

The parameters were the number of cluster heads and cluster sizes. The simulation revealed that OCTOPUS generated fewer cluster heads than the probabilistic technique. Figure 8.8 shows that the number of cluster heads created by the probabilistic approach usually exceeds OCTOPUS's number. This is a result of the creation of more redundant cluster heads in a cluster since many cluster head candidates satisfied the probability condition. This problem was encountered in OCTOPUS to a lesser degree.

The experiment showed that the introduction of extra variables in addition to the node degree increased the pool of cluster head candidates. The second parameter to be evaluated related directly to the number of cluster heads. An increase in cluster numbers leads to smaller cluster size. The cluster size generated by OCTOPUS was often higher than the cluster size created by the probabilistic approach as depicted in Figure 8.9. In general, OCTOPUS tends to optimize by accommodating the maximum number of nodes under the minimum number of clusters.

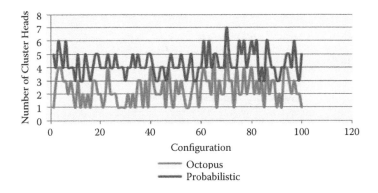

FIGURE 8.8 Number of cluster heads generated by OCTOPUS and probabilistic techniques.

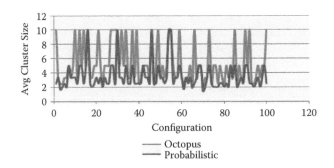

FIGURE 8.9 Cluster size generated by OCTOPUS and probabilistic techniques.

8.7   CONCLUSIONS AND DISCUSSION

This chapter proposes a distributed technique to build a communication topology for mobile sensor networks. The approach starts with clustering nodes into separate groups. Communication among nodes of the same cluster is facilitated by an elected cluster head. Communication with other clusters is performed by a selected gateway node. The technique produces a communication map that specifies a role for each node in a network.

The evaluation of the technique revealed performance comparable with other well known approaches. The technique has many applications. With on-demand operation, a node can request a reconfiguration if a failure occurs at cluster head or gateway level. Similarly, a node joining a network can request a role through reconfiguration. The approach can be employed as a scheduled technique for reorganization and reconfiguration or an alternative to topology control and maintenance. However, the cost of frequent updates and reevaluation of the topology should be considered.

ACKNOWLEDGMENTS

The authors gratefully acknowledge the support received from NASA through a grant titled "NNX10AC25A: Guided Planetary Surface Exploration Using Rovers and Wireless Sensor Networks," and from the International Information Systems Security Certification Consortium (ISC²).

REFERENCES

1. T. Caimu and C. S. Raghavendra, "Energy Efficient Adaptation of Multicast Protocols in Power Controlled Wireless Ad Hoc Networks," *Mobile Netw. Appl.*, vol. 9, pp. 311–317, Aug 2004.
2. D. Li, H. Du, W. Chen, "Power Conservation for Strongly Connected Topology Control," *Wireless Sensor Networks*, pp. 9–16, 2008.
3. J. Ma, M. Gao, Q. Zhang and M. L. Ni, "Energy-Efficient Localized Topology Control Algorithms," *IEEE Trans. Parallel Distrib. Syst.*, vol. 18, pp. 711–720, May 2007.
4. C. Xiuzhen, M. Ding, D. H. Du and X. Jia, "Virtual Backbone Construction in Multi-Hop Ad Hoc Wireless Networks," *Wireless Commun. Mobile Comput.*, vol. 6, pp. 183–190, March 2006.
5. V. Gayathri, E. Sabu and T. Srikanthan, "Size-restricted Cluster Formation and Cluster Maintenance Technique for Mobile Ad Hoc Networks," *Int. J. Netw. Manag.*, vol. 17, pp. 171–194, March 2007.
6. M. Yu, J. H. Li, and R. Levy, "Mobility Resistant Clustering in Multi-Hop Wireless Network," *J. Netw.*, vol.1, pp.12–19, May 2006.
7. F. G. Nocetti, J. S. Gonzales and I. Stojmenovic, "Connectivity Based K-hop Clustering in Wireless Networks," in *Proc. 35ᵗʰ Annu. Hawaii Int. Conf. Syst. Sci.*, 2002, vol. 7, pp. 188.
8. J. H. Li, M. Yu, and R. Levy, "Distributed Efficient Clustering Approach for Ad Hoc and Sensor Networks," in *Proc. 1st Int. Conf. Mobile Ad hoc and Sensor Netw.*, Dec. 2005.
9. L. V. Fausett, *Fundamentals of Neural Networks Architectures: Algorithms and Application.* New Delhi, India: Pearson Education, 2005.
10. D. J. Baker and A. Ephremides, "A Distributed Algorithm for Organizing Mobile Radio Telecommunication Network," *in Proc. 2nd Int. conf. Distrib. Comput. Syst.*, Apr. 1981, pp. 476–483.

11. M. Gerla and J.T.C. Tsai, "Multicluster, Mobile, Multimedia Radio Network," *Wireless Networks*, Vol. 1, pp. 255–265, 1995.

12. S. Basagni, "Distributed and Mobility-Adaptive Clustering for Multimedia Support in Multi-Hop Wireless Networks," *in Proc. Veh. Technol. Conf.,* Sept. 1999, vol. 2, pp. 889–893.

13. H. Tan, W. Zeng and L. Bao, "PATM: Priority-Based Adaptive Topology Management for Efficient Routing in Ad Hoc Networks," in *Proc. Int. Conf. Comput. Sci.,* 2005, pp. 485–492.

14. W.R. Heinzelman, A. Chandrakasan and H. Balakrishnan, "Energy-Efficient Communication Protocol for Wireless Microsensor Networks," in *Proc. 33rd Hawaii Int. Conf. Syst. Sci.,* 2000, vol. 8, pp. 8020.

15. T. W. Haynes, S. T. Hedetniemi and P.J. Slater, *Fundamentals of Domination in Graphs.* New York: Marcel Dekker, 1998.

16. T.W. Haynes, S. T. Hedetniemi and P.J. Slater, *Domination in Graphs Advanced Topics.* New York: Marcel Dekker, 1998.

17. M. Chatterjee, S.K. Das and D. Turgut, D., "WCA: A Weighted Clustering Algorithm for Mobile Ad Hoc Networks," *Cluster Comput. J.,* vol. 5, no 2, pp. 193–204, Apr. 2002.

18. P. Wan, K. Alzoubi, and O. Frieder, "Distributed Construction of Connected Dominating Set in Wireless Ad Hoc Networks," in *Proc. INFOCOM,* Apr. 2004, vol. 9, pp. 141–149.

19. T. T. Wu and K. F. Su, "Determining Active Sensor Nodes for Complete Coverage without Location Information," *Int. J. Ad Hoc Ubiquitous Comput.,* vol. 1, pp. 38–46, Nov. 2005.

20. H. Lui and R. Gupta, "Selective Backbone Construction for Topology Control," in *Proc. 1st IEEE Int. Conf. Mobile Ad-hoc and Sensor Syst. (MASS),* Oct. 2004, pp. 41–50.

21. L. Ruan, D.H. Du, X. Jia, W. Wu, Y. Li and K.-I Ko, "A Greedy Approximation for Minimum Connected Dominating Sets," *Theoret. Comput. Sci.,* vol. 339, pp. 325–230, Dec. 2004.

22. M. Min, C.X. Huang, S. C.-H. Huang, W. Wu, H. Du, and X. Jia, "Improving Construction of Connected Dominating Set with Steiner Tree," *Wireless Sensor Netw.,* vol. 35, pp. 111–119, May 2006.

23. K.M. Alzoubi, P.-J. Wan and O. Frieder, "Distributed Heuristics for Connected Dominating Sets in Wireless Ad Hoc Networks," *J. Commun. Netw.,* vol. 4, pp. 24–29, Mar. 2002.

24. X. Cheng, X. Huang, D. Li, W. Wu and D.-Z. Du, "Polynomial-Time Approximation Scheme for Minimum Connected Dominating Set in Ad Hoc Wireless Networks," *Networks,* vol. 42, pp. 202–208, Sep. 2003.

25. C. H. Papadimitriou, "On the Complexity of Integer Programming," *J. ACM,* vol. 28, pp. 765–768, Oct. 1981.

26. P. Ratanchandani and R. Kravets, "A Hybrid Approach to Internet Connectivity for Mobile Ad Hoc Networks," in *Proc. of WCNC,* Mar. 2003, vol. 23, pp. 1522–1527.

27. P.-M. Ruiz, and A.-F. Gomez-Skarmeta, "Adaptive Gateway Discovery Mechanisms to Enhance Internet Connectivity for Mobile Ad Hoc Networks," *Ad Hoc and Sensor Wireless Netw.,* vol. 1, 159–177, Mar. 2005.

28. L. R. Foulds, *Optimization Techniques, An Introduction.* New York: Springer, 1981.

29. L. A. Wolsey, *Integer Programming.* New York: Wiley-Interscience, Sep. 1998.

30. G. Sierksma, G. A. Tijssen and P. V. Dam, *Linear and Integer Programming: Theory and Practice.* New York: Marcel Dekker, Mar. 1996.

31. T. Chen, H. Zhang, X. F. Zhou, G. M. Maggio, and I. Chlamtac, "CogMesh: A Cluster Based Cognitive Radio Mesh Network," in *Proc. 2nd IEEE Int. Symp. New Frontiers Dynam. Spectrum Access Netw.,* Apr. 2007, pp. 168–178.

32. *Statistics Toolbox User's Guide.* Natick, MA: MathWorks Inc., 2009, MATLAB Tutorials, Version 7.

33. L. Barolli, H. Ando, F. Xhafa, A. Durresi, R. Miho and A. Koyama, "Evaluation of an Intelligent Fuzzy-based Cluster Head Selection System for WSNs using Different Parameters," *AINA workshops,* Mar. 2011.

34. Q. Zhang and D. Agrawal, "Dynamic Probabilistic Broadcasting in Manets," *J. Par. Dist. Comput.,* Feb. 2005.

9 Network Structure for Delay-Aware Applications in Wireless Sensor Networks

Chi-Tsun Cheng, Chi K. Tse, and Francis C.M. Lau

CONTENTS

9.1 INTRODUCTION

Wireless sensor networks comprise large numbers of wireless sensor nodes. The compact design of a wireless sensor node allows it to fit into most environments and perform close-range sensing. Such a unique feature makes wireless sensor networks highly suitable for monitoring in extreme conditions.

In most sensing applications, sensing information should be sampled regularly and spatially over time. Wireless sensor nodes are battery-powered devices. To maintain required sensing coverage over such a period, wireless sensor nodes should conserve their energy aggressively but remain operational. Much prior work has focused on conserving energy by clustering.

A network with clustering is divided into several clusters. Within each cluster, one of the sensor nodes is elected as a *cluster head* (CH) and the rest serve as *cluster members* (CMs). The CH collects data from its CMs directly or in a multi-hop manner. By organizing wireless sensor nodes into clusters, energy dissipation is reduced by decreasing the number of nodes involved in long distance transmission [1].

The number of data transmissions and energy consumption levels can be reduced further by performing data and decision fusion on nodes along the data aggregation

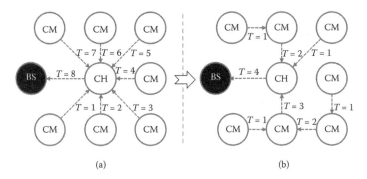

FIGURE 9.1 (a) Data collection in a two-hop network. (b) Data collection in improved multi-hop network. Circles with CMs represent cluster members. Circles with CHs represent cluster heads. Filled circles with BS notations represent base stations. A dashed arrow represents a data link and arrow direction indicates direction of data flow.

path. Clustering provides a significant improvement in energy saving. In sensor networks with clusters, it is common for a CH to collect data from its CMs one by one.

Let T be the average transmission delay among nodes. Data packets generated by sensor nodes are considered highly correlated, and thus a node is always capable of fusing all received packets into a single packet by means of data or decision fusion techniques [2,3]. Referring to Figure 9.1a, a *base station* (BS) will take $8 \times T$ to collect a complete set of data from the network. By transforming the network into a multi-hop system as shown in Figure 9.1b, the time needed by the BS to collect a full set of data from the network can be reduced to $4 \times T$. In a modified network, in addition to a shorter delay in data collection, CMs need smaller buffers to handle incoming data while waiting for the relevant CH to become available.

The aim of this chapter is to investigate the characteristics of delay-aware data collection network structures in wireless sensor networks. An algorithm for forming such a network structure is proposed for different scenarios. The proposed algorithm operates between the data link layer and the network layer. The algorithm will form networks with minimum delays in data collection and will also try to keep the transmission distances among wireless sensor nodes at low values to limit the energy consumed in communications.

Section 9.2 briefly reviews related work. Section 9.3 defines the proposed network structure. Section 9.4 explains the algorithm for forming the proposed network structure in different scenarios. A numerical analysis in Section 9.5 demonstrates how different network structures perform in terms of delays in data collection. Simulation results and their analyses are covered in Sections 9.6 and 9.7, respectively. Finally, the chapter concludes with Section 9.8.

9.2 RELATED WORK

Due to the energy constraints of individual sensor nodes, energy conservation has become a major issue in sensor networks. In wireless sensor networks, a large portion of the energy in a node is consumed in wireless communications. The amount of energy consumed in a transmission is proportional to the corresponding communication distance, and long distance communications between nodes and the BS are usually not encouraged.

One way to reduce energy consumption in sensor networks is to adopt a clustering algorithm [1] for organizing sensor nodes into clusters. Within each cluster, one node is elected as the CH. The CH is responsible for (1) collecting data from its CMs, (2) fusing the data via data and decision fusion techniques, and (3) reporting the fused data to the remote BS. In each cluster, the CH is the only node involved in long distance communications. Energy consumption of the whole network is therefore reduced.

Intensive research [2–5] focused on reducing energy consumption by forming clusters with appropriate network structures. Heinzelman et al. proposed a clustering algorithm called LEACH [2]. In networks using LEACH, sensor nodes are organized in multiple-cluster two-hop (MC2H) networks (CMs → CH → BS). Using clustering allows the number of long distance transmissions to be reduced greatly.

Lindsey and Raghavendra proposed another clustering algorithm called PEGASIS [3] that operates differently by organizing sensor nodes into a single-chain (SC) network. A single node on a chain is selected as the CH. By minimizing the number of CHs, the energy consumed in long distance transmission is further minimized.

Tan and Körpeoğlu developed PEDAP [4] based on the idea of a minimum spanning tree (MST). Besides minimizing the number of long distance transmissions, the communication distances among sensor nodes are minimized.

Fonseca et al. [5] proposed the collection tree protocol (CTP)—a kind of gradient-based routing protocol that uses *expected transmissions* (ETX) as routing gradients. ETX indicates the number of expected transmissions of a packet necessary for it to be received without error [6]. Paths with low ETX are expected to have high throughput. Nodes in a network using CTP will always pick a route with the lowest ETX. In general, the ETX of a path is proportional to the corresponding path length [7].

Thus, CTP can greatly reduce the communication distances among sensor nodes. All these algorithms show promising results in energy saving. However, a network formed by an energy-efficient clustering algorithm may not necessarily be desirable for data collection. An analysis of the performances of these network structures in terms of data collection efficiency will be covered in Section 9.5.

The focus of this chapter is investigation of the data collection efficiency of networks formed by different clustering algorithms. Event-triggering algorithms such as TEEN [8] and APTEEN [9] will not be considered in this chapter. Related work on data collection efficiency was done by Florens et al. [10]. They derived lower bounds on data collection time for various network structures. However, the effect of data fusion—one of the major features of sensor networks—was not considered. Wang et al. [11] proposed link scheduling algorithms for wireless sensor networks that raised network throughput considerably. However, their work assumed that data links among wireless sensor nodes were predefined.

In contrast, the objective of this chapter is to form data links among wireless sensor nodes and thus to shorten the delays in data collection. A related work by Solis and Obraczka [12] studied the impact of timing in data aggregation for sensor networks. Chen and Wang [13] investigated the effects of network capacity under different network structures and routing strategies. A similar study was done by Song and He [14]. They defined capacity as the maximum end-to-end traffic a network can handle. A delay in data collection was not their major concern.

9.3 PROPOSED NETWORK STRUCTURE

The proposed network structure is a tree. To deliver maximum data collection efficiency, the number of nodes N in the proposed network structure must be restricted to $N = 2^p$ where $p = 1,2,\ldots$. We will show later that such a restriction can be relaxed by foregoing some performance.

Each CM will be ranked as an integer between 1 and p. A node with rank k will form $k - 1$ data links with $k - 1$ nodes. The $k - 1$ nodes are assigned different ranks starting from $1, 2,\ldots$ up to $k - 1$. All these $k - 1$ nodes will become the child nodes of the node with rank k. The node with rank k will form a data link with a node of a higher rank. The higher rank node will become the parent of the node with rank k.

The CH is considered a special case. The CH has the highest rank in the network. Instead of forming a data link with a node of higher rank, the CH will form a data link with the BS. By following this logic, the distribution of ranks will follow an inverse exponential base-2 function, as shown in Table 9.1. An example of the proposed network with $N = 16$ is shown in Figure 9.2. The example utilizes $5 \times T$ for the BS to collect all data from 16 nodes. By dividing the time domain into time slots of duration T, the above process will require five time slots.

Lemma 1

Consider a network with $N = 2^p$ where $p = 1,2,\ldots$. Data packets generated by sensor nodes are considered highly correlated, and thus a node is always capable of fusing all received packets into a single packet by means of data and decision fusion techniques. By adopting the proposed network structure, a node i of rank $k \geq 2$ (where $k \in Z$) requires $k - 1$ time slots to collect data from all its child nodes.

Proof: Consider a network with $N = 2^p$ where $p = 1,2,\ldots$. For a node of rank $k = 2$, the time slots required for it to collect data from all its child nodes equals the number of child nodes it has (1). Thus the case for $k = 2$ is true. Now let us assume that any node of connection $k = n$ requires $n - 1$ time slots to collect all data from its child nodes. Node

---

TABLE 9.1
CM Rank Distributions in Proposed Network Structure[a]

Rank	1	2	...	$\log_2 N - 1$	$\log_2 N$
Number of Nodes	$\dfrac{N}{2^1}$	$\dfrac{N}{2^2}$	...	$\dfrac{N}{2^{(\log_2 N - 1)}}$	$\dfrac{N}{2^{(\log_2 N)}}$

[a] Network size $N = 2^p$ where $p = 1,2,\ldots$.

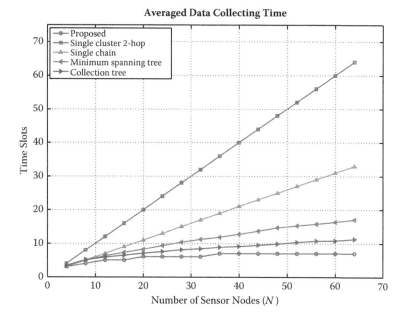

FIGURE 9.2 Proposed network structure with network size N = 16. Circles with CMs represent cluster members. The circle with CH represents the cluster head. The filled circle (BS) represents the base station. Rank of each node is represented by the variable k. A dashed arrow represents a a data link and arrow direction indicates direction of data flow.

i of rank $k = n + 1$ has n directly connected child nodes. Each child node has a different rank ranging from 1 to n. Thus, the child nodes need 0 to $n - 1$ time slots to collect data from all their subchild nodes plus one extra time slot to report their aggregated data to node i. Therefore, the maximum time slots required for node i to collect data from all its child nodes is $k - 1 = n$. By induction, the Lemma is proved. ∎

Theorem 1

Consider a network with $N = 2^p$, where $p = 1,2,\ldots$. Data packets generated by sensor nodes are considered highly correlated, and thus a node is always capable of fusing all received packets into a single packet by means of data and decision fusion. By adopting the proposed network structure, the number of time slots $t(N)$ required for the BS to collect data from the whole network is given by:

$$t(N) = \log_2 N + 1. \tag{9.1}$$

Proof: Consider a network with $N = 2^p$, where $p = 1,2,\ldots$. By adopting the proposed network structure, the CH is the only node with the highest ranking:

$$k_{max} = \log_2 N + 1.$$

From Lemma 1, the number of time slots $t(N)$ required for a CH, with rank k_{max}, to collect data from all its child nodes is:

$$t(N) = k_{max} - 1 = \log_2 N.$$

Thus, the number of time slots $t(N)$ required for the BS to collect data from the whole network is the time slots required by the CH to collect data from all its child nodes plus one:

$$t(N) = \log_2 N + 1. \qquad \blacksquare$$

9.4 NETWORK FORMATION ALGORITHM

The last section demonstrated that delays in data collection by a wireless sensor network can be reduced greatly by adopting the proposed network structure. Since energy consumption is always a major issue in the study of wireless sensor networks, the objective of the proposed network formation algorithms is retaining the network structure while keeping data collection energy consumption at a low value.

A wireless sensor node can be considered a device consisting of three major components: the microcontroller unit (MCU), transceiver unit (TCR), and sensor board (SB). Each unit will consume a certain amount of energy while operating. The energy consumed by a wireless sensor node i can be expressed as:

$$E_{i\_SN} = E_{i\_MCU} + E_{i\_TCR} + E_{i\_SB} \qquad (9.2)$$

where E_{i_MCU} represents the energy consumed by the MCU, E_{i_TCR} represents the energy consumed by the TCR, and E_{i_SB} represents the energy consumed by the SB. E_{i_TCR} can be further expressed as:

$$E_{i\_TCR} = E_{i\_TCR\_RX} + E_{i\_TCR\_TX}(d_i) \qquad (9.3)$$

where $E_{i\_TCR\_RX}$ denotes the energy consumed by the TCR in receiving mode, while $E_{i\_TCR\_TX}(d_i)$ denotes the energy consumed by the TCR to transmit a distance of d_i. The total energy consumed by a network of N sensor nodes is expressed as:

$$E_{TOT}(N) = \sum_{i=1}^{N} \left(E_{i\_MCU} + E_{i\_TCR\_RX} + E_{i\_TCR\_TX}(d_i) + E_{i\_SB} \right). \qquad (9.4)$$

Normally, E_{i_MCU}, E_{i_TCR}, and $E_{i\_TCR\_RX}$ are constants. On the other hand, $E_{i\_TCR\_TX}(d_i)$ is a function of d_i and depends heavily on the network structure. Therefore, Equation (9.4) can be simplified as follows:

$$E_{TOT}(N) = c_1 + \sum_{i=1}^{N} E_{i\_TCR\_TX}(d_i) \qquad (9.5)$$

where C_1 is a constant. Assuming that the path loss exponent equals 2, $E_{i\_TCR\_TX}(d_i)$ can be further expressed as:

$$E_{i\_TCR\_TX}(d_i) = E_{i\_TCR\_EC} + E_{i\_TCR\_PA} d_i^2 \qquad (9.6)$$

where $E_{i\_TCR\_EC}$ is the energy consumed by the TCR's electronic circuitry, while $E_{i\_TCR\_PA}$ denotes the energy consumed by the power amplifier of the TCR. Both $E_{i\_TCR\_EC}$ and $E_{i\_TCR\_PA}$ are constants and therefore Equation (9.5) can be expressed as:

$$E_{TOT}(N) = c_1 + c_2 + c_3 \sum_{i=1}^{N} d_i^2 \qquad (9.7)$$

where C_2 and C_3 are constants. Equation (9.7) shows that the total energy consumption of a network can be minimized by reducing $\sum_{i=1}^{N} d_i^2$. Thus, the objective of the proposed network formation algorithms is to construct a proposed network structure while keeping $\sum_{i=1}^{N} d_i^2$ at low value. In this section, a network formation algorithm is proposed to achieve this objective.

Basically, the operation of the proposed network formation algorithm is to join clusters of the same size together. It can be implemented in a centralized or decentralized fashion. Specifically, the decentralized version can be described as follows.

Each node is labeled with a unique identity and marked as level w. The unique identity serves only as an identification that has no relation to sensor node locations or connections. Here w is a function that represents the number of nodes in a cluster. A cluster of i nodes has a w value equal to $\log_2 i$. Since nodes are disconnected initially (no data link exists among wireless sensor nodes), the N nodes can be considered N level 0 clusters. Within each cluster, one node will be elected as the subCH. We denote SCH(w) as a subCH of a level w cluster. In the proposed algorithm, one SCH can make connection (set up a data link) only with another SCH of the same level. Because each cluster contains only one node, all nodes begin as SCH(0). The dimensions of the terrain (t_x, t_y) are provided to the sensor nodes before deployment.

Each SCH performs random back-off and then broadcasts a *density probing packet* (DPP) to neighboring SCHs within a distance of $r_{dp} = (t_x^2 + t_y^2)^{\frac{1}{2}}$ m. Note that a DPP is much smaller than a data packet. A SCH can use the number of received DPPs together with the dimensions of the terrain to estimate the total number of nodes (N_{est}) in the network and use the N_{est} to adjust its communication distance r_{com}. A definition of r_{com} will be covered later in this section.

Each SCH will do a random back-off and then broadcast an *invitation packet* (IVP) to its neighbors within r_{com} m. The IVP contains the level w and the identity of the issuing SCH. A SCH will estimate the distances to its neighboring SCHs using the signal strength of the IVPs received. It will also count the number of IVPs received. If the number of IVPs exceeds a predefined threshold or a maximum duration is reached, a SCH will send a *connection request* (CR) to the nearest neighbor. If both SCHs are nearest neighbors of each other, a connection will be formed between them.

After they are connected, the two SCHs and their level w clusters will form a composite level $w + 1$ cluster. One of the two involved SCHs will become the chief

SCH of the composite cluster and will listen to the communication channel and reply to CRs from lower levels with a *rejecting packet* (RP). When no more CRs from lower levels can be heard, the chief SCH will start to connect with other SCHs of the same level.

If a RP is received, a SCH will send a CR to its next nearest neighbor in its database. If no such exists, the SCH will increase its r_{com} and then broadcast a CR using the new r_{com}. Upon receiving the CR, a SCH of the same level will grant the request if it is still waiting for a CR.

If no connection can be made within a certain period, all neighbors of the same level are unavailable, or all CRs have been rejected, the SCH will increase its r_{com} and broadcast the CR again. This process repeats as long as $r_{com} < \sqrt{t_x^2 + t_y^2}$. If $r_{com} = \sqrt{t_x^2 + t_y^2}$, the SCH will make connection with the BS directly.

The above processes continue until no more connections can be formed.

In the proposed algorithm, the communication distance r_{com} is defined as:

$$r_{com} = \frac{\sqrt{t_x^2 + t_y^2}}{\alpha - \beta - w}, \quad \beta + w < \alpha. \tag{9.8}$$

β is a constant which is set to 0 initially. Parameter α is the estimated maximum rank of a node in the network, expressed as:

$$\alpha = \lceil \log_2(N_{est}) \rceil + 1. \tag{9.9}$$

Initially, all SCHs are with $w = 0$ and $\beta = 0$ and therefore start broadcasting their IVPs with $r_{com} = \left(\sqrt{t_x^2 + t_y^2}\right)(\alpha - 0 - 0)^{-1}$. If a SCH has made a connection with another SCH, its level will be increased by 1 ($w = 1$). After that, the chief SCH of the composite cluster will broadcast its IVP with $r_{com} = \left(\sqrt{t_x^2 + t_y^2}\right)(\alpha - 0 - 0)^{-1}$. The r_{com} is designed to be increased with w because when SCHs are paired to form composite clusters, the average separation among composite clusters increase. It is more energy efficient to start broadcasting with a longer communication range.

However, if no connection can be made, a SCH will increase its β by 1. This will increase r_{com}, which can facilitate the search for available SCHs. A SCH will increase its r_{com} by incrementing β until a connection can be made. The sum of β and w is defined to be smaller than α to ensure r_{com} is upper bounded by the diagonal of the sensing terrain.

In step 3 of the proposed algorithm, a SCH will send a *connection request* (CR) to its nearest neighbor if the number of received IVPs exceeds $\aleph$. Here $\aleph$ is the expected number of IVPs to be received, expressed as:

$$\aleph = \left\lceil \frac{\eta \pi r_{com}^2 - 1}{w} \right\rceil. \tag{9.10}$$

Parameter η is the density of the network that can be estimated using the N_{est} obtained earlier.

When implemented in a decentralized control manner, the proposed algorithm may end with multiple composite clusters if the number of nodes is not equal to 2^p where $p = 1,2....$ SCHs of these composite clusters will communicate with the BS directly. By pairing up composite clusters of same sizes, the algorithm will end with composite clusters of completely different sizes. Considering the BS as the root of the network, the number of time slots required by the BS to collect data from all sensor nodes is:

$$t(N) = \lceil \log_2(N+1) \rceil. \tag{9.11}$$

In contrast, the proposed algorithm can also be carried out at the BS as a centralized control algorithm. The BS is assumed to have the coordinates of all sensor nodes in the network. When the number of nodes is not equal to 2^p, where $p = 1,2...$, dummy nodes can be virtually added to the calculation process, depending on the application. If a single cluster is required, dummy nodes should be virtually added to fulfill the requirement of $N = 2^p$, where $p = 1,2....$

Dummy nodes are not physical; they are virtual nodes used to facilitate computation at the BS. If multiple clusters can be formed, dummy nodes are not essential. When dummy nodes are virtually added, they will have infinite separations with the real nodes and with themselves. Note that whenever a real SCH connects with a dummy SCH, the real one will always be the chief of the composite cluster. This is to ensure that the removal of dummy nodes at the end of the calculation process will not partition the network. The number of time slots required by the BS to collect data from all sensor nodes will be governed by Equations (9.1) [realWOdummy] for single clusters and (9.11) for multiple clusters.

9.5   NUMERICAL ANALYSES

Theorem 2

Assume that each sensor node can communicate only with one sensor node at a time and that data fusion is applicable. For a single cluster network of $N = 2^p$ nodes where $p = 1,2,...$, the minimum number of time slots required by the BS to collect data from N sensor nodes is:

$$t(N)_{min} = \log_2 N + 1. \tag{9.12}$$

Proof: Given a period of t time slots, a parent node v can collect data from at most t directly connected child nodes provided the t child nodes use different time slots to communicate. Within these t child nodes, the uth node will report data at time slot u, which implies the uth node can collect data from at most $u - 1$ directly connected child nodes of itself before it has to report data to its parent node.

Therefore, for a period of t time slots, a parent node v can receive data from at most 2^t nodes (including itself). On the other hand, the minimum number of time slots required for a parent node to collect data from N nodes (including itself) is $\log_2 N$. Thus, the minimum number of time slots required for a BS to collect data from N nodes is $t(N)_{min} = \log_2 N + 1$.

For a single cluster network with N nodes, where $N > 0 \mid N \in Z$, the minimum number of time slots required for a BS to collect data from N nodes is:

$$t(N)_{min} = \lceil \log_2 N \rceil + 1. \tag{9.13}$$

From Theorem 1 and Equation (9.1), we see that the proposed network structure is an optimum structure in terms of data collection efficiency provided that:

1. Each sensor node can communicate only with one sensor node at a time.
2. Data fusion can be carried out at every sensor node.
3. Sensor nodes belong to a single cluster with a single CH.

The same idea can be applied to a multiple cluster network by considering the BS as the root of the network structure. The minimum number of time slots required for a BS to collect data from N nodes is:

$$t(N)_{min} = \lceil \log_2(N + 1) \rceil. \tag{9.14}$$

Using Equation (9.11), it can be shown that the proposed network structure is again an optimum structure in terms of data collection efficiency provided that:

1. Each sensor node can communicate only with one sensor node at a time.
2. Data fusion can be carried out at every sensor node.
3. The network consists of multiple clusters.

In a MC2H network with N nodes organized in g clusters where $N \geq \sum_{m=1}^{g} m$, the time slots required by the BS to collect data from all sensor nodes is minimized when all clusters have different numbers of nodes. Therefore, each cluster can communicate with the BS in an interleaved manner. Meanwhile, the number of nodes in the largest cluster should be minimized so that the total number of time slots required by the BS is also minimized. An example for $g = 2$ is shown below.

Example 1

For a MC2H network of N nodes organized in 2 clusters (where $N \geq 3$), to achieve maximum data collection efficiency, the number of nodes in these two clusters should be equal to:

$$\frac{N+1}{2} \text{ and } \frac{N-1}{2}, \quad \text{for } N \text{ odd}$$

$$\frac{N}{2} + 1 \text{ and } \frac{N}{2} - 1, \quad \text{for } N \text{ even} \tag{9.15}$$

The minimum number of time slots $t(N)_{min}$ required by the BS to collect data from all sensor nodes equals the number of nodes in the largest cluster. Therefore:

$$t(N)_{min} = \begin{cases} \dfrac{N+1}{2}, & N \text{ is odd} \\ \dfrac{N}{2}+1, & N \text{ is even.} \end{cases} \tag{9.16}$$

In general, for a MC2H network of N nodes organized in g clusters where $N \geq \sum_{m=1}^{g} m$, the number of nodes in the jth cluster can be written as:

$$\left\lfloor \frac{N - S_g + (j-1)(g+1)}{g} \right\rfloor + 1, \; j = 1, 2, 3, \ldots, g \tag{9.17}$$

where $\lfloor u \rfloor$ denotes the nearest integer smaller than u and $S_g \geq \sum_{m=1}^{g} m$. Thus, the minimum number of time slots $t(N)_{min}$ required by the BS to collect data from all sensor nodes equals:

$$t(N)_{min} = \left\lfloor \frac{N - S_g + (g-1)(g+1)}{g} \right\rfloor + 1. \tag{9.18}$$

Based on Equation (9.17), the optimum number of clusters g_{opt} for a MC2H network in terms of data collection efficiency can be obtained from the following inequality:

$$\frac{(1+g)g}{2} \geq N$$

$$\Rightarrow \quad g \geq \frac{-1 + \sqrt{1^2 + 8N}}{2} \tag{9.19}$$

$$\Rightarrow \quad g_{opt} = \left\lceil \frac{-1 + \sqrt{1^2 + 8N}}{2} \right\rceil \tag{9.20}$$

where N is the number of nodes in the network and g is the number of clusters.

Theorem 3

For a MC2H network of N nodes organized in g clusters of completely different sizes where $g \leq N < S_g$, the minimum time slots $t(N)_{min}$ required by the BS to collect data from all sensor nodes is equal to the number of clusters in the system, i.e., g.

Proof: Consider an extreme case. A MC2H network of N nodes is organized in g clusters of completely different sizes where $N = S_g$. All the g clusters will have different numbers of nodes ranging from 1 to g. The minimum number of time slots $t(N)_{min}$ required by the BS to collect data from all sensor nodes is equal to the number of nodes in the largest cluster (g).

Suppose one node has to be removed from the network such that N is reduced to $N - 1$. To maintain the number of clusters in the network, this particular node must be removed from one of the clusters except the one with a single node. Removing a node from any of the clusters will cause two clusters to have the same number of nodes. During a data collection process, the 2 clusters of the same size will have to do interleaving, which will not affect $t(N)_{min}$. Therefore, the minimum number of time slots $t(N)_{min}$ required by the BS is always equal to the number of clusters in the system.

In contrast, for a MC2H network with N nodes organized in g clusters, where $N \geq g$, the number of time slots required by the BS to collect data from all sensor nodes is maximized when $N - (g - 1)$ nodes belong to the same cluster. The remaining $g - 1$ clusters will all have a size of 1, and we have:

$$t(N)_{max} = \begin{cases} N - (g-1), & N > 2(g-1) \\ N - (g-1) + 1, & N = 2(g-1) \\ g, & \text{otherwise.} \end{cases} \tag{9.21}$$

In a SC network, the number of time slots required by the BS to collect data from all sensor nodes is minimized when the CH is at the middle of the chain:

$$t(N)_{min} = \begin{cases} \dfrac{N+1}{2} + 1, & N \text{ is odd} \\ \dfrac{N}{2}, & N \text{ is even} \end{cases} \tag{9.22}$$

where N is the number of nodes in the network. Conversely, in a SC network, the number of time slots required by the BS to collect data from all sensor nodes is maximized when the CH is at the end of the chain:

$$t(N)_{max} = N \tag{9.23}$$

where N is the number of nodes in the network.

In networks using MST and CTP, the number of time slots required by the BS to collect data from all sensor nodes is lower bounded by Equations (9.12) and (9.13). Conversely, the number of time slots required by the BS to collect data from all sensor nodes is maximized when the resultant networks of MST and CTP are in single cluster two-hop structures upper bounded by $t(N)_{max} = N$. ■

9.6 SIMULATIONS

In this section, the proposed network structure will be compared with MC2H, a SC, MST, and CTP networks. The networks having N nodes (N varies from 4 to 64) with step size of 4 will be distributed randomly and evenly on a sensing field of $50 \times 50 \ m^2$. The center of the sensing field is located at $(x, y) = (25 \ m, 25 \ m)$.

In the simulations, synchronization among wireless sensor nodes is maintained by the physical layer and the data link layer. Wireless sensor nodes are assumed to be equipped with CDMA-based transceivers [15]. Interference due to parallel transmissions can be alleviated by utilizing different spreading sequences in different data links.

Media access control during network formation is handled by the MAC sublayer and assumed to be satisfactory. A node can receive or transmit at any time. In the simulation, a wireless sensor node is always capable of fusing all received packets into a single packet by means of data and decision fusion. The size of an aggregated packet is independent of the number of packets received. For each network, the averaged data collection time (DCT) will be used to indicate data collection efficiency. The communication distance of a network is represented by the following function:

$$\psi = \sum_{i=1}^{N} u_i d_{i\_B}^h + \sum_{i=1}^{N-1} \sum_{j=i+1}^{N} C_{ij} d_{ij}^h \tag{9.24}$$

where u_i indicates CHs ($u_i = 1$) and CMs ($u_i = 0$). Parameter d_{i_B} is the distance between a CH and the BS. Here c_{ij} is an indicator of the presence ($c_{ij} = 1$) or absence ($c_{ij} = 0$) of a data link between node i and node j. Furthermore, d_{ij} is the geographical distance between nodes i and j. In the simulations, the path loss exponent h is assumed to be 2. The BS is assumed to be at the center of the sensing field ($x = 25$ m, $y = 25$ m).

For the simulations on network lifetime, each node received 50 J of energy. The energy model of the wireless sensor nodes is the same as the one introduced in Section 9.4. A network will perform data collection periodically. Its lifetime is defined as the number of data collection processes (rounds) that it can accomplish before any of its nodes runs out of energy.

Each data packet is p_{data} bits long. Other packets are all regarded as controls. Each control packet is p_{ctrl} bits long. Values of the parameters used in the energy model are shown in Table 9.2. The network structures under test are classified as Type I (single cluster network structure) and Type II (multiple cluster network structure). Under this classification, all structures under test belong to Type I, whereas MC2H and the proposed network structure belong to both types. Because the numbers of

TABLE 9.2

Values of Parameters Used in Simulations

Parameter	Value
$E_{i\_TCR\_RX}$	50×10^{-6} J/bit
$E_{i\_TCR\_EC}$	50×10^{-6} J/bit
$E_{i\_TCR\_PA}$	100×10^{-9} J/bit/m^2
E_{i_MCU}	5×10^{-6} J/bit
E_{i_SB}	50×10^{-6} J/bit
p_{data}	1024 bits
p_{ctrl}	64 bits

clusters in Type I and Type II network structures are different, results obtained for different structure types should not be compared directly.

For the proposed network structure to work as Type I, sufficient dummy nodes are added. To work as Type II, the proposed network structure can be constructed without adding dummy nodes. The cluster number of the MC2H network is fixed to 1 when it works as a Type I structure. Here, ψ of the MC2H network is minimized by selecting the node with minimum separations from its fellow nodes as the CH. To work as a Type II structure, CHs in the MC2H networks are selected randomly [2], while the optimum number of CHs is selected according to Equation (9.20). In both configurations, CMs are connected to their nearest CHs.

The SC network can only work as a Type I structure, and the chain is formed by using a greedy algorithm [3]. To minimize DCT, the node closest to the middle of the chain (in terms of hops) will be selected as the CH. Similar to the SC network, the MST network can work only as a Type I structure. Networks will be formed by using Prim's algorithm [4]. To minimize DCT, the node with the smallest separation (in terms of hops) from all leaf nodes will be selected as the CH.

In networks using CTP, the node closest to the center of the sensing terrain serves as the root of the collection tree. The ETX of a path is expressed as the squared value of the path length [6]. The root of the tree will have an ETX of 0. The ETX of an arbitrary node is its cumulative ETX through its parent nodes to the root [5]. Each node will choose its best route by selecting the path with the minimum accumulated ETX. Simulation results are shown in Figures 9.3 through 9.8. All results presented in this chapter were averaged over 100 simulations.

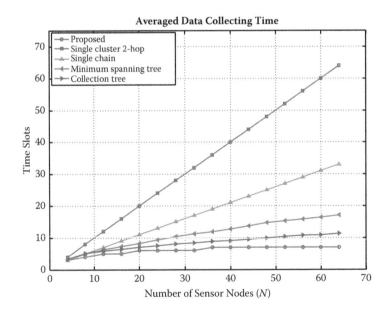

FIGURE 9.3 Averaged data collection times of different single tree structures.

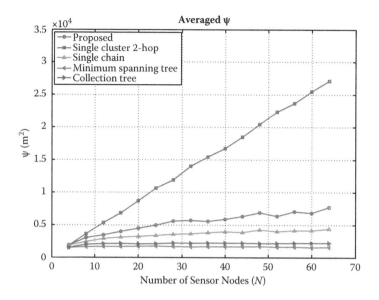

FIGURE 9.4   Averaged ψ values of different single tree structures.

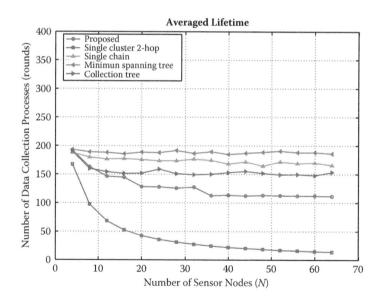

FIGURE 9.5   Averaged lifetimes of different single tree structures.

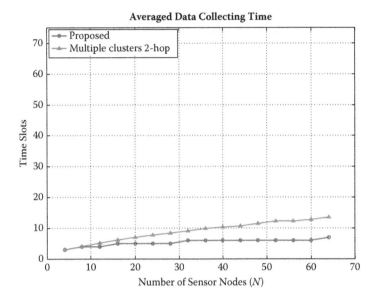

FIGURE 9.6 Averaged data collection times of different multiple-cluster structures.

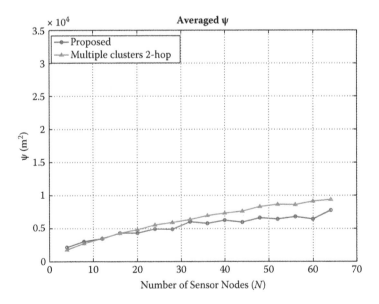

FIGURE 9.7 Averaged ψ values of different multiple-cluster structures.

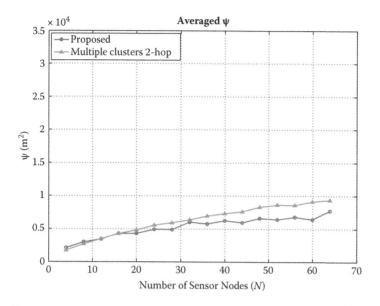

FIGURE 9.8 Averaged lifetimes of different multiple-cluster structures.

9.7 ANALYSIS

As expected from Section 9.5 the DCTs of networks with the proposed network structure were the lowest among Type I structures. In simulations among the six Type I structures, DCTs of networks with the proposed network structure were the lowest, followed by networks with CTP. Since the aim of the MST is to minimize the total weight of edges, it does not perform well in reducing bottlenecks and therefore ranks fourth.

In a SC network, it takes a very long time for data to propagate from both ends of the chain to the CH at the middle. This explains why SC networks have much higher DCTs than networks with the proposed network structure. With the single cluster two-hop structure, the networks with MC2H do not present any advantage in reducing DCT. The MC2H network is the one with the highest DCT among Type I structures.

For minimizing ψ, the MST network clearly ranks first. The ETX used in networks with CTP can greatly reduce the communication distances among sensor nodes and thus they rank second. Nodes in networks with SC structure try to reduce the total communication distance by connecting to their nearest neighbors only. The strategy is effective for networks with small numbers of nodes. However, to maintain a single chain structure, it is unavoidable for the SC network to increase its ψ as the number of nodes increases. The SC network therefore ranks third.

Using the optimization techniques employed in Section 9.4, the ψ of networks constructed using the proposed algorithm does not increase drastically as N increases and it ranked fourth. With all sensor nodes connected to a single CH, the MC2H network is the structure with the highest ψ without question.

In a data collection process, a node with connection degree k is required to receive $k - 1$ data packets, perform $k - 1$ times of data fusion, and transmit one data packet. A node with a high connection degree will certainly consume more energy that one with a low degree. Therefore, a network that has a uniform connection degree distribution is more likely to yield a longer lifetime than one with a non-uniform distribution. The minimum spanning property of a MST network can evenly distribute the connection degree among nodes. Therefore, the MST network can achieve the highest network lifetime for networks with $N > 4$.

In a SC network, most of the nodes have connection degrees equal to 2. The energy consumed by each node in receiving and fusing data is relatively low. Therefore, the SC network ranks second. Similar to the simulations on the ψ, network lifetime of networks with CTP and the proposed network structure rank third and fourth, respectively. In a SC2H network, all CMs are connected to a single CH. The CH is heavily loaded and has a very high energy consumption. This explains why a SC2H network has the lowest lifetime among all network structures under test.

In simulations of the two Type II structures, networks formed by the proposed algorithm exhibited the lowest DCTs. Although the networks with MC2H structures were tuned to give the optimum number of clusters, there was no control of the distribution of sensor nodes in each cluster. The DCTs of networks with MC2H structures therefore greatly increased as N increased.

For the same reason, network lifetimes of networks with MC2H structures decreased gradually as N increased. In terms of ψ, both the proposed and the MC2H network structures yielded similar results when $N \leq 12$. For $N > 12$, networks constructed by the proposed algorithm revealed lower ψ than MC2H structures.

The gap increases further as N increases. According to Equation (9.20), a MC2H network is most efficient if it has g_{opt} clusters where g_{opt} is proportional to N. As N increases, g_{opt} increases and thus more nodes are involved in long distance transmissions. This also happens to networks constructed by the proposed algorithm. Nevertheless, due to the special topology of the proposed network structure, the number of clusters increases at a lower rate.

This explains why ψ of a network formed by the proposed algorithm is lower than those constructed in MC2H structures. This also explains why networks formed by the proposed algorithm can achieve longer network lifetimes than those with MC2H structures.

9.8 CONCLUSIONS

In this chapter, a delay-aware data collection network structure and its formation algorithm are proposed. To handle different applications, a network formation algorithm may be implemented in a centralized or decentralized manner.

The performance of the proposed network structure was compared with a multiple-cluster two-hop network structure, a single-chain network structure, a minimum spanning tree network structure, and a collection tree network structure. The proposed structure exhibited the most efficient data collection time among all the network structures mentioned. The proposed structure can greatly reduce data collection time while maintaining total communication distance and network lifetime at acceptable values.

REFERENCES

1. J.N. Al-karaki and A.E. Kamal. 2004. Routing techniques in wireless sensor networks: a survey. *IEEE Wireless Communications*, 11, 6–28.
2. W.B. Heinzelman, A.P. Chandrakasan, and H. Balakrishnan. 2002. An application-specific protocol architecture for wireless microsensor networks. *IEEE Transactions on Wireless Communications*, 1, 660–670.
3. S. Lindsey and C.S. Raghavendra. 2002. PEGASIS: power-efficient gathering in sensor information systems. In *Proceedings of IEEE Conference on Aerospace*, pp. 1125–1130.
4. H.Ö. Tan and Í. Körpeoĝlu. 2003. Power-efficient data gathering and aggregation in wireless sensor networks. *ACM SIGMOD Record*, 32, 66–71.
5. R. Fonseca, O. Gnawali, K. Jamieson et al. 2007. The collection tree protocol. *Tiny OS Enhancement Proposals*, 123, December.
6. D.S. Couto. 2004. High-throughput routing for multi-hop wireless networks. Ph.D. Dissertation. Massachusetts Institute of Technology, Cambridge.
7. O. Tekdas, J.H. Lim, A. Terzis et al. 2009. Using mobile robots to harvest data from sensor fields. *IEEE Wireless Communications*, 16, 22–28.
8. A. Manjeshwar and D.P. Agrawal. 2001. TEEN: a routing protocol for enhanced efficiency in wireless sensor networks. In *Proceedings of 15th International Symposium on. Parallel and Distributed Processing*, pp. 2009–2015.
9. A. Manjeshwar and D.P. Agrawal. 2002. APTEEN: a hybrid protocol for efficient routing and comprehensive information retrieval in wireless sensor networks. In *Proceedings of 16th International Symposium on Parallel and Distributed Processing*, pp. 195–202.
10. C. Florens, M. Franceschetti, and R. J. McEliece. 2004. Lower bounds on data collection time in sensory networks. *IEEE Journal of Selected Areas in Communications*, 22, 1110–1120.
11. W. Wang, Y. Wang, X.Y. Li et al. 2006. Efficient interference-aware TDMA link scheduling for static wireless networks. In *Proceedings of 12th Annual International Conference on Mobile Computing and Networking*, pp. 262–273.
12. I. Solis and K. Obraczka. 2004. The impact of timing in data aggregation for sensor networks. In *Proceedings of IEEE International Conference on Communications*, Vol. 6, pp. 3640–3645.
13. Z.Y. Chen and X.F. Wang. 2006. Effects of network structure and routing strategy on network capacity. *Physics Reviews E*, 73, 1–5.
14. M. Song and B. He. 2007. Capacity analysis for flat and clustered wireless sensor networks. In *Proceedings of International Conference on Wireless Algorithms, Systems and Applications*, pp. 249–253.
15. Texas Instruments. n.d. CC2520. http://focus.ti.com/lit/ds/symlink/cc2520.pdf

10 Game-Theoretic Models for Camera Selection in Video Network

Yiming Li and Bir Bhanu

CONTENTS

10.1 INTRODUCTION

Video networking requires two capabilities: camera handoff and camera selection. Camera handoff is a dynamic process by which the system transfers the tracking of an object from one camera to another without losing the object in the network. It is clear that a manual camera handoff will become unmanageable for a large number of cameras. Furthermore, it is unrealistic to display and manually monitor surveillance videos captured from a large number of cameras simultaneously. Therefore, video surveillance systems need another capability: camera selection.

Traditionally, these two components were treated as a single problem handled on the borders of the camera FOVs only. In this chapter, we provide a new perspective on camera selection and handoff based on game theory. We define camera selection as a camera–object map that indicates at each time instant which camera is used to follow which object, allowing us to handoff from one camera to another when necessary and using the best available camera.

When multiple cameras are used for tracking and "see" the same object, the algorithm can automatically provide *optimal* and *stable* solutions of camera assignment. Since the game-theoretic approaches can deal with multiple criteria optimization, we can choose the best camera based on multiple criteria selected a priori.

10.2 RELATED WORK AND OUR CONTRIBUTIONS

10.2.1 RELATED WORK

Many papers have discussed approaches to camera selection in a video network. The traditional approaches generally fall into topology-based and statistics-based categories. The topology-based approaches [1–4] rely on geometrical relationships among cameras. These relationships become very complicated when the topology becomes complex and it is difficult to learn the topology based on random traffic patterns [5]. The approaches based on statistics [6–10] usually depend on object trajectories. Other factors such as orientation, shape, and face, although important for visual surveillance, are not considered. A comparison of our approach and related work is summarized in Table 10.1.

10.2.2 OUR CONTRIBUTIONS

Our approach differs from the conventional approaches [1–4,6–8,12,14], shown in Table 10.1 in certain key steps: (1) proposal of game-theoretic approach; (2) development of multiple criteria for tracking; (3) evaluation of proposed approach using real data; and (4) revealing promising results.

10.3 GAME-THEORETIC APPROACHES

10.3.1 WHY GAME-THEORETIC APPROACHES?

Game theory is well known for analyzing interactions and conflicts among multiple agents [15,16]. Analogously, in a video sensor network, collaborations and competitions among cameras exist simultaneously. The solution lies in the collaboration of all the available cameras that can "see" the target person and track him or her as long as possible. Conversely, the available cameras also compete for the right to track the person so that they can maximize their own utility.

This conflict allows us to view camera assignment in a game-theoretic manner. A game is an interactive process [17] among all the participants or players who

TABLE 10.1
Comparison of Our Work and Related Work

Author	Approach	Comments
Javed et al. [1]	Find limits of overlapping FOVs of multiple cameras. Cameras' choices are based on distance between person and edge of FOV.	Significant increase in computation with increases of cameras and persons.
Park et al. [2]	Create distributed look-up tables based on how well cameras image specific location.	Based only on single criterion; depends on data that may not be available for each network.
Jo and Han [3]	Construct handoff function by computing COR for selected pairs of points in FOVs of two cameras.	Computation cost is high for increased handoff resolution. Handoff ambiguity and failure arise from large number of cameras or persons.
Qureshi and Terzopoulos [4]	Model conflicts may arise during camera assignment to satisfy constraints. Each camera assignment that passes hard constraints is assigned a weight. Camera with highest weight is selected.	Leader node required. All solutions that can pass constraints must be calculated and best one selected. Ranking becomes more time consuming as problem becomes more complex.
Kettnaker and Zabih [6]	Choose object trajectory with highest posterior probability.	Camera assignment based on paths of objects. Frontal view can be lost even when available.
Chang and Gong [7]	Use Bayesian modality fusion. Bayesian network used to combine multiple modalities for matching subjects among multiple cameras.	Geometry- or recognition-based. Hard to combine them for decision; focuses on occlusion, not other criteria for assignment.
Kim and Kim [8]	Find dominant camera for an object based on ratio of numbers of blocks on object to ratio in camera FOV.	Dominant camera must be aided by views from other cameras. Only size and angle are considered.
Cai and Aggarwal [12]	Matches human objects by using multivariate Gaussian models.	Camera calibration needed.
Tessens et al. [14]	Messages related to distributed process go to base station to determine principal camera based on score calculated by experiments offline.	Uses distributed and centralized control. Criteria determined offline so method not suitable for active camera selection and control. Principal view complemented by helper views.
Our work	Models camera assignment as potential game using vehicle–target model and uses bargaining mechanism to obtain converged solution with few iterations.	Can decide camera assignment based on multiple criteria; no camera calibration is needed; independent of camera network topology.

strive to maximize their utilities. The utility of a player refers to the benefit accrued from the game. In our problem, a multi-player game exists for each person to be tracked. The available cameras are the players. If a system involves multiple persons, the result is a multiple of multi-player games played simultaneously [18].

10.3.2 Potential Game Model

We will introduce a vehicle–target model first in this section. Vehicle–target assignment [13] is a classical multi-player *potential game* intended to allocate a set of vehicles to a group of targets to achieve optimal assignment. By viewing the persons tracked as "vehicles" and the cameras as "targets," we can adopt the vehicle–target assignment model to choose the best camera for each person. In the next section, a non-potential game—weakly acyclic—model will be proposed.

10.3.2.1 Problem Formulation as Potential Game

We are concerned with three utilities for this model: (1) global utility is the overall degree of satisfaction for tracking performance; (2) camera utility measures how well a camera tracks assigned persons; and (3) Person utility indicates how well a person is satisfied during tracking by some camera. Our objective was to maximize global utility while ensuring that each person was tracked by the best camera. When competing with other available cameras, cameras "bargain" with each other. Finally an assignment decision is made based on a set of probabilities.

Figure 10.1 illustrates an overview of the approach. Moving objects are detected in multiple video streams. Their properties (size of the minimum bounding rectangle and other region properties such as color, shape, and location within FOV) are computed. Various utilities are calculated based on user-supplied criteria and bargaining processes among available cameras are executed based on the prediction of person utilities from the previous iteration step.

The results obtained from the strategy execution are in turn used to update the camera utilities and the person utilities until the strategies converge. Finally, the cameras with the most converged probabilities are used for tracking. This assignment of persons to the best cameras leads to the solution of the handoff problem in multiple video streams.

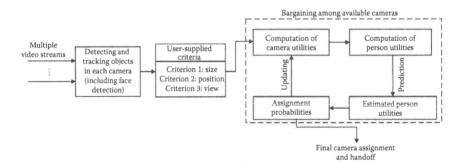

FIGURE 10.1 Overview of potential game approach.

10.3.2.2 Computation of Utilities

We define the following properties for the potential game model of our system:

1. A person P_i can be in the FOVs of multiple cameras. The available cameras for P_i belong to set A_i. C_0 is a virtual camera that does not actually exist. We assume a virtual camera C_0 is assigned to P_i when no real camera in the network is available to track P_i.
2. A person can be assigned to only one camera. The assigned camera for P_i is named a_i.
3. Each camera can be used to track multiple persons.

We use a to denote the camera assignment for all the persons, and a_i denotes the assigned camera for P_i. For P_i, when we change the camera assignment to a_i'' while assignments for other persons remain the same, if we have

$$U_{P_i}\left(a_i', a_{-i}\right) < U_{P_i}\left(a_i'', a_{-i}\right) \Leftrightarrow U_g\left(a_i', a_{-i}\right) < U_g\left(a_i'', a_{-i}\right) \tag{10.1}$$

the person utility U_{P_i} is said to be aligned with the global utility U_g, where a_{-i} denotes the assignments for persons other than P_i, i.e., $a_{-i} = (a_1, \ldots, a_{i-1}, a_{i+1}, \ldots, a_{N_P})$. The camera assignment result a can also be expressed as $a = (a_i, a_{-i})$. In cases where this alignment property is met, we call this method a potential game. In our circumstances, we define global utility as:

$$U_g(a) = \sum_{C_j \in C} U_{C_j}(a) \tag{10.2}$$

where $U_{C_j}(a)$ is the camera utility to be generated by all the engagements of persons with a particular camera C_j. Now, we define the person utility:

$$U_{P_i}(a) = U_g\left(a_i, a_{-i}\right) - U_g\left(C_0, a_{-i}\right) = U_{C_j}\left(a_i, a_{-i}\right) - U_{C_j}\left(C_0, a_{-i}\right) \tag{10.3}$$

where, C_0 is a virtual camera. The person utility $U_{P_i}(a)$ can be viewed as a marginal contribution of P_i to the global utility. To calculate Equation (10.3), we must construct a scheme to calculate camera utility $U_{C_j}(a)$. We assume N_{Crt} criteria can evaluate the quality of a camera used to track an object. Thus, camera utility can be built as:

$$U_{C_j}\left(a_i, a_{-i}\right) = \sum_{s=1}^{n_P} \sum_{l=1}^{N_{Crt}} Crt_{sl} \tag{10.4}$$

where n_P is the number of persons currently assigned to camera C_j for tracking and Crt constitutes criteria supplied by the user. Plugging Equation (10.4) into (10.3) we obtain:

$$U_{P_i}\left(a_i, a_{-i}\right) = \sum_{l=1}^{N_{Crt}} \left(\sum_{s=1}^{n_P} Crt_{sl} - \sum_{\substack{s=1 \\ s \neq P_i}}^{n_P} Crt_{sl} \right) \tag{10.5}$$

where $s \neq P_i$ means that we exclude person P_i from the those tracked by Camera C_j. Notice that we must normalize criteria when designing them. The types of criteria to be fed into the bargaining mechanism do not matter, as discussed below.

The choice of criteria to be used for camera assignment and handoff depends on user requirements. Different criteria may be required for different applications such as power consumption, time delay, image resolution, and other factors [18,25,29]. Camera assignment results may change when different criteria are applied. Our goal was to find a proper camera assignment solution quickly based on various criteria supplied by users.

10.3.2.3 Bargaining among Cameras

As stated previously, our goal was to optimize person utility and global utility. Competition among cameras leads to the Nash equilibrium [21] as a solution to camera assignment and handoff. Unfortunately, the Nash equilibrium may not be unique. Some of the solutions may not be stable and stability is desired. To solve this problem, a bargaining mechanism among cameras was devised to make them reach a compromise and generate a stable solution.

When bargaining, the assignment in the kth step is based on a set of probabilities $p_i(k) = \left[p_i^1(k), ..., p_i^l(k), ..., p_i^{n_C}(k) \right]^T$ where n_C is the number of cameras that can "see" the person P_i and $\sum_{l=1}^{n_C} p_i^l(k) = 1$, with each $0 \leq p_i^l(k) \leq 1$, $l = 1, ..., n_C$. We can generalize $p_i(k)$ as $p_i(k) = \left[p_i^1(k), ..., p_i^l(k), ..., p_i^{N_C}(k) \right]^T$.

At each bargaining step, we assign a person to the camera with the highest probability. We assume that one camera has no information about other cameras' utilities at the current step and this makes calculation of all possible current person utilities difficult. This led us to develop the concept of predicted person utility $\bar{U}_{P_i}(k)$. Before we decide the final assignment profile, we predict the person utility using the previous person's utility information in the bargaining steps.

As shown in Equation (10.5), person utility depends on camera utility and we thus predict the person utility for every possible camera that may be assigned to track it. Each element in $\bar{U}_{P_i}(k)$ is calculated by:

$$\bar{U}_{P_i}^l(k+1) = \begin{cases} \bar{U}_{P_i}^l(k) + \dfrac{1}{p_i^l(k)}\left(U_{P_i}(a(k)) - \bar{U}_{P_i}^l(k)\right), & a_i(k) = A_i^l \\ \bar{U}_{P_i}^l(k), & otherwise \end{cases} \tag{10.6}$$

with the initial state $\bar{U}_{P_i}^l(1)$ assigned arbitrarily as long as it is within a reasonable range for $\bar{U}_{P_i}(k)$ for $l = 1,..., n_C$. After these predicted person utilities are calculated, it can be proved that the equilibrium for the strategies lies in the probability distribution that maximizes its perturbed predicted utility [10]:

$$p_i(k)' \bar{U}_{P_i}(k) + \tau H(p_i(k)) \tag{10.7}$$

where $H\left(p_i(k)\right) = -p_i(k)' log(p_i(k))$ is the entropy function and τ is a positive parameter belonging to [0,1] that controls the extent of randomization and *log* means taking the log of every element of the column vector $p_i(k)$, yielding a column vector. The larger τ is, the faster the bargaining process converges; the smaller the τ, the more accurate the result. A tradeoff is required in selecting the value of τ. After several steps of calculation, the result of $p_i(k)$ tends to converge. Thus, we finally get a stable solution that is at least suboptimal [13].

10.3.3 WEAKLY ACYCLIC GAME MODEL

The weakly acyclic game belongs to a super class of potential games and thus can relax some of the limitations of a potential game model. In a weakly acyclic game, the local utility function does not have to be aligned with a global utility. This provides much flexibility in utility function design.

In the previous section, we applied a potential game model to the camera selection and handoff problem. The same utility designs were used for all the cameras aligned with the global utility function. However, in this weakly acyclic game model (as shown in the next section), a video network may have different utility functions for cameras in different groups. For example, in one group of cameras, face resolution may be the most important factor Another group of camera may consider tracking of individuals more important. Thus, the assumption that only one camera can be used to track one object is also relaxed. Furthermore, this flexibility means we did not have to calculate global utility and the system could be distributed to a large extent.

10.3.3.1 Problem Formulation as Weakly Acyclic Game

We view the actions that a camera can take as strategies a player can use in a game. The final solution to the camera selection and handoff problem is a set of strategies for each camera. Figure 10.2 presents an overview of the proposed utility-based weakly acyclic game approach for camera selection and handoff.

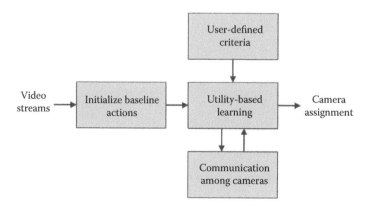

FIGURE 10.2 Overview of weakly acyclic game approach.

Let B_j denote the set of actions for camera j, then $B_j = \{$sleep, awake and free of task, awake and recording, awake and recording for $P_1, \ldots$ awake and recording for $P_i, \ldots$, awake and recording for $P_{N_p}\}$. At each time instant, the actual action of camera j, P_j, may equal any of the elements in set B_j. The camera assignment profile is the set of each camera's strategy at the current moment $b = \{b_1, b_2, \ldots, b_j, \ldots, b_{nC-1}, b_{nC}\}$ where a_j is the strategy for camera j.

Let $a_{-j} = \{b_1, b_2, \ldots, b_{j-1}, b_{j+1}, \ldots, b_{nC}\}$ be the strategy profile for all the cameras other than j. A game G is said to be weakly acyclic if there exists a better reply path (the system will gain more utility by taking strategies along this path) starting at some strategy profile and ending at some pure Nash equilibrium of G. A better reply path is a sequence of camera action profiles $b(1)$, $b(2)$, $\ldots$, $b(T)$ such that for each successive pair $(b(t), b(t + 1))$ there is exactly one player such that $b(t) \neq b_j(t + 1)$ and for that player $U_{C_j}(b(t+1)) > U_{C_j}(b(t))$ [23].

Since we can change the camera actions at each iteration of the learning process, we can always find a better reply path for camera action assignment until it reaches the optimal Nash equilibrium. Thus this process can be modeled as a weakly acyclic game. Since the game is finite and the path cannot cycle back on itself according to the definition, the last reply path must be the optimal solution of the camera assignment.

10.3.3.2 Utility-Based Learning Algorithm

First, we group our cameras into four categories: G1 = indoor; G2 = court yard; G3 = corridor; and G4 = border. Figure 10.3 shows a map of the cameras used in the experiments. The border cameras are always awake. If an object is detected by a border camera, it will awaken its neighboring cameras and we then run the utility-based learning algorithm [25] for all awake cameras.

The general idea of the learning algorithm is to first randomly choose a baseline action for each awake camera. At each learning iteration, we update the camera

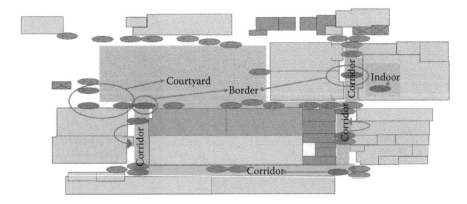

FIGURE 10.3 Map of VideoWeb camera network. The encircled cameras were used in our experiments. The disks that are not in circles are the cameras not used. The dark areas indicate regions where the experiments were carried out. (*Source:* H. Nguyen et al. 2009. In *Proceedings of ACM/ IEEE ICDSC*. With permission.)

actions with an exploration rate ε that is actually a probability. If the result is a better reply path, we replace the baseline actions with the better path. Otherwise, we keep the baseline actions the same as in the previous step.

A better reply path in our problem is a set of camera action assignments that can give a camera that is changing its strategy a greater utility level than it had during the previous iteration step. By following this procedure for each iteration, we will finally achieve a better reply path that ensures maximum utility for each camera.

The exploration rate ε is chosen as any real number belonging to [0, 1]. The trade-off for choosing ε is that when a large ε is used, the learning process will converge quickly and impose the risk of losing the optimal result. Conversely, if ε is small enough, we can guarantee that this learning process will access the optimal Nash equilibrium with arbitrarily high probability [23].

10.4   EXPERIMENTS

10.4.1   EXPERIMENTAL DATA

We tested two proposed game-theoretic approaches on five cases as shown in Table 10.2. We used Axis 215 PTZ cameras placed arbitrarily in all our experiments. To fully test whether the proposed approach would select the best camera based on user-supplied criteria, some of the FOVs of the cameras were allowed to interact and some were non-overlapping. The experiments were conducted without previous camera calibration. The trajectories were chosen randomly by the walking persons, and we empirically assigned values to the parameters [18,25].

10.4.2   RESULTS OF POTENTIAL GAME MODEL APPROACH

10.4.2.1   Performance Metrics

The intent was to track walking persons seamlessly by having the system follow a person as long as he or she appeared in the FOV of at least one camera. In cases where multiple cameras could "see" a person, we assumed that the camera that could

---

TABLE 10.2
Cases for Experiments

Case Number	N_p	N_c	Id	Cd	Od	N_f
1	2	3	√			450
3	5	3			√	1016
4	6	3		√	√	1329
5	8	4		√	√	1728
6	8	6	√	√	√	2238

N_p = number of persons; N_c = number of cameras; Id = indoor; Cd = corridor; Od = outdoor; N_f = number of frames.

"see" the person's face was preferred. This is because the frontal view of a person can provide more interesting information than other views generated by a surveillance system. Based on this criterion, we defined a camera assignment error as: (1) failing to track a person (some cameras see the person but no single camera tracks him or her) or (2) failing to obtain a frontal view of a person when possible.

10.4.2.2 Evaluation of Potential Game Model Approach

Experiment #1: Comparison of potential game and COR approaches ($N_C = 3$, $N_P = 2$, indoor)—We compared the proposed potential game model with the co-occurrence-to-occurrence ratio (COR) approach proposed by Jo and Han [3]. In the COR approach, the mean probability that a moving object is detected at a location x in the FOV of a camera is called an occurrence at x. The mean probability that moving objects are simultaneously detected at x in the FOV of one camera and x' in the FOV of another camera is called a co-occurrence of x and x'.

The COR approach decides whether two points are in correspondence by calculating the COR. If the COR is higher than some predefined threshold, the two points are considered in correspondence with each other. When one point approaches the edge of the FOV of one camera, the system will hand off to another camera that has its corresponding point. However, the COR approach in Reference [3] utilized only two cameras.

We generalized this approach to cases with more cameras by comparing the accumulated CORs in the FOVs of multiple cameras. We randomly selected 100 points on a detected person, trained the system for ten frames to construct the correspondence for these 100 points, calculated the cumulative CORs in the FOVs of different cameras, and selected the one with the highest value for handoff.

Figure 10.4 shows error frames with the COR approach. Note that the COR approach can only switch a camera to another one when the person is about to leave the FOV. COR cannot select the best camera based on other criteria. The number of handoffs by our approach is larger than that for COR (Table 10.3). Using the

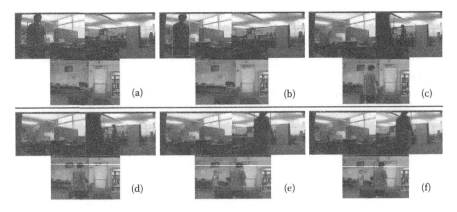

FIGURE 10.4 Experiment #1. Some camera hand-off errors arose from co-occurrence-to-occurrence ratio (COR) approach in a three-camera, two-person case.

TABLE 10.3

Comparison of Error Rates for COR and Proposed Approaches

		Experiment #2	
	Approach	Handoffs	Error Rate (%)
Case 1	COR	5	45.62
	Potential game approach	8	5.56

definitions of error from Section 10.4.2.1, the error rates for these two cases are compared in the table. The COR approach lost the frontal views more easily. Examples are Figure 10.4b (person), Figure 10.4d (frontal view of person), and Figure 10.4f (frontal view of person).

10.4.3   Results of Weakly Acyclic Game Model Approach

10.4.3.1   Performance Metrics

In the weakly acyclic game model, we eliminated the alignment requirement and this allowed us to study more applications than those introduced in the previous section. This section covers experiments for two different application scenarios:

- Application 1: tracking people in a camera network smoothly; frontal face preferred
- Application 2: tracking people in network smoothly; frontal face preferred only in selected area

Accordingly, we used different error definitions for different applications. Err_1 and Err_2 denote errors for the above three applications, respectively. In Application 1, we assign an error if a person is in the coverage of the camera network but no camera is selected to follow him or her. We also want to follow a person smoothly so that assignment does not dither among multiple cameras. We also consider more than three handoffs in ten frames (1 second) an error. Because frontal face view is preferred, we consider the failure of camera C_j to track a person whose frontal face can be seen as another error.

For Application 2, we assigned some cameras in a group to look for frontal views of persons in a selected area. Other cameras in other groups will focus on smooth tracking of persons.

Experiment #2: People across camera network, frontal face preferred, Application 1—Users prefer a frontal view when it is available. We tested the proposed approach using Cases 2, 3, and 4. We also compared our results with the potential game approach. To achieve the bargaining process in the potential game approach, we had to calculate utilities for both persons and cameras. This almost doubled the computational cost compared to the cost of weakly acyclic game approach where only camera utility is required.

Figure 10.5 shows sample frames for comparison. Note that the limited number of iterations used still allowed the proposed approach to find the favorable camera in most cases. The potential game approach may have lost the best camera because of the lack of available iterations. For example, in Figure 10.5a (Case 2, Frame 426), the proposed approach selected Camera 2 for the person since his frontal view is available. The potential game approach lost the frontal view and selected Camera 1. Similarly, in Case 2 Frame 574, the proposed approach successfully selected the frontal view for the person while the potential game approach lost it. For a better comparison, Figure 10.6 plots the final utilities obtained for Camera 0 in each

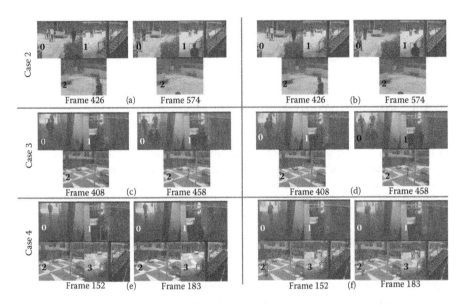

FIGURE 10.5 Experiment #2. Example frames for Cases 2 through 4 for comparison of the weakly acyclic game approach (left column) and potential game approach (right column). These are some of the example frames in which the proposed weakly acyclic game approach could pick the camera showing the front face of a person. The potential game approach failed to do so. The number is the camera identification.

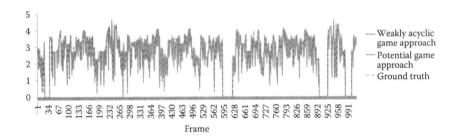

FIGURE 10.6 Experiment #2. Camera 0's utility in each frame is shown for the weakly acyclic game approach and the potential game approach.

TABLE 10.4

Comparison of Proposed Weakly Acyclic Game and Potential Game Approaches

Case	Approach	Handoffs	Dithering	Error Rate (%)
Case 2	Weakly acyclic game	172	5	9.85
	Potential game	196	35	15.64
Case 3	Weakly acyclic game	199	7	11.05
	Potential game	239	39	19.23
Case 4	Weakly acyclic game	221	7	11.92
	Potential game	267	48	22.56

frame for Case 2 by using the proposed approach and the potential game approach, respectively.

In Table 10.4, we summarize the results for both approaches. A weakly acyclic game also involves fewer handoffs because it considers smoothness when switching among cameras. The weakly acyclic approach also has a lower error rate, especially in more complicated cases, because it can reach an optimal solution with fewer iterations. The experiments with the particle filter tracker displayed a higher error rate than those with annotated videos because the camera utilities based on tracking quality are related to tracking errors.

Experiment #3: Tracking people across camera network (different groups of cameras have different criteria), Application 2—We used two cameras at the court yard, three at the corridor, and another indoor camera. We only acquired frontal views from the indoor camera. The other cameras were for smooth tracking and handoffs only. Unlike the previous experiments using cameras within the same group, this experiment involved four groups of cameras. We also considered time delay when the cameras were awakened by the border cameras in this experiment.

To demonstrate the advantage of the proposed approach, we compared the potential game approach, a non-game-theoretic approach [24], and a fuzzy-based approach. In Reference [24], each candidate camera had two states for an object its FOV: (1) a non-selected state and (2) a selected state for tracking. Camera handoff was based on the camera's previous state S_i and the tracking level state SS_i defined by estimating the position measurement error in the monitoring area. The tracking level state can be decided by the proposed Criterion 2, S_{i2}.

Figure 10.7 shows example frames from different approaches. We obtained similar results from the three approaches when a person was visible in only one camera. For example, in Frames 239, 406, and 712, all three approaches selected Camera 0 for indoor persons. However, if more than one camera was available to track a person, the fuzzy-based approach tended to hand off to another camera only when the person was about to leave the FOV of the current camera; the fuzzy method did not consider the view or size of the person.

Frame 576 is an example of this situation. Although the potential game approach performed better than the fuzzy-based approach in this sense, it sometimes failed

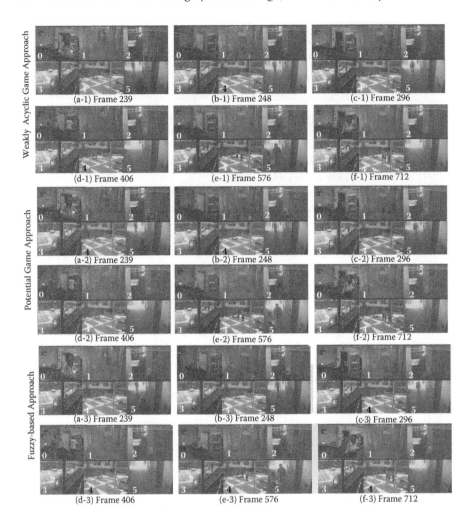

FIGURE 10.7 Experiment #3. Comparison of example frames with different approaches in Case 6. The number is the camera identification. Cameras 1 and 5 served as border cameras. Cameras 0, 1, 2, and 5 handled frontal views. Cameras 3 and 4 were for smooth tracking.

to hand off to the best camera because of the limited iterations (up to 37 iterations were required in this case to get the best camera). The fuzzy approach also dithers among cameras a lot because smoothness is not considered.

The final results are compared in Table 10.5. Note that the error rate for the fuzzy-based approach is high even though it does not have a dithering problem. This is because the approach does not take into account the view of a person. Thus, by applying the error definition in Section 10.4.3.1, the fuzzy-based approach exhibited a very high error rate for the camera that should have picked up a frontal view.

We noted a higher error rate for the weakly acyclic approach. When time delay is considered, the optimal camera may be missed for one of two frames when awakened

TABLE 10.5
Comparison of Proposed and Other Approaches

Case	Approach	Handoffs	Dithering	Error Rate (%)
Case 5	Weakly acyclic game	322	9	11.64
	Potential game [8]	418	52	25.64
	Fuzzy-based [17]	283	7	29.96

by one or more border cameras. In the transit frame, the time delay criterion may have a 0 value and lead to a low payoff, causing the camera not to be selected by the system. The error rate caused by the time delay was 1.23%.

10.5   CONCLUSIONS

In this chapter, we proposed two game-theoretic methods, the potential game and weakly acyclic game approaches, to the camera assignment and handoff problem. We developed a set of intuitive criteria and compared them with each other and in combination. Our experiments showed that the combined criterion performed the best based on the error definitions provided in Section 10.4.3.1.

In the potential game model, since the utility inputs of the bargaining process largely depend on user-supplied criteria, our proposed approach can be task-oriented. Unlike the conventional approaches that perform camera handoffs only when an object enters or leaves FOVs, we can select the best camera based on predefined criteria.

As the complexity of a scenario (number of people and cameras) increases, the number of iterations needed for the weakly acyclic game approach is fewer than the number required for the potential game approach. Thus, the weakly acyclic game approach is much more efficient than the potential game approach. The key merit of the proposed approaches is the use of a theoretically sound game theory framework with bargaining mechanisms and utility-based learning algorithms for camera assignment in a video network and thus obtain a stable solution with a reasonably small number of iterations.

REFERENCES

1. O. Javed, S. Khan, Z. Rasheed et al. 2000. Camera hand-off: tracking multiple uncalibrated stationary cameras. In *Proceedings of IEEE Workshop on Human Motion*, pp. 113–118.
2. J. Park, P. C. Bhat, and A. C. Kak. 2006. A look-up table-based approach for solving the camera selection problem in large camera networks. *International Workshop on Distributed Smart Cameras*.
3. Y. Jo and J. Han. 2006. A new approach to camera hand-off without camera calibration for the general scene with non-planar ground. In *Proceedings of ACM International Workshop on Video Surveillance and Sensor Networks*, pp. 195–202.
4. F. Z. Qureshi and D. Terzopoulos. 2008 Multi-camera control through constraint satisfaction for pedestrian surveillance. In *Proceedings of AVSS*, pp. 211–218.

5. X. Zou and B. Bhanu. 2008. Anomalous activity classification in the distributed camera network. In *Proceedings of ICIP*, pp. 781–784.
6. V. Kettnaker and R. Zabih. 1999. Bayesian multi-camera surveillance. In *Proceedings of CVPR*, vol. 2, pp. 253–259.
7. T. Chang and S. Gong. 2001. Bayesian modality fusion for tracking multiple people with a multi-camera system. *European Workshop on Advanced Video-based Surveillance Systems*.
8. J. Kim and D. Kim. Probabilistic Camera Hand-off for Visual Surveillance. *2008. ICDSC*.
9. J. Kang, I. Cohen, and G. Medioni. 2003. Continuous tracking within and across camera streams. In *Proceedings of CVPR*, pp. 267–272.
10. O. Javed, K. Shafique, and M. Shah. 2005. Appearance modeling for tracking in multiple non-overlapping cameras. In *Proceedings of CVPR*, pp. 26–33.
11. B. Song and A. Roy-Chowdhury. 2007. Stochastic adaptive tracking in a camera network. *ICCV*.
12. Q. Cai and J.K. Aggarwal. 1999. Tracking human motion in structured environments using a distributed camera system. In *Proceedings of PAMI*, pp. 1241–1247.
13. G. Arslan, J.R. Marden and J.S. Shamma. 2007. Autonomous vehicle–target assignment: a game-theoretical formulation. *ASME Transactions on Dynamic Systems, Measurement, and Control*, 129, 584–596.
14. L. Tessens et al. 2008. Principal view determination for camera selection in distributed smart camera networks. *ICDSC*.
15. R.B. Myerson. 1991. *Game Theory: Analysis of Conflict*. Cambridge, MA: Harvard University Press.
16. M.J. Osborne. 2003. *An Introduction to Game Theory*. Oxford, U.K.: Oxford University Press.
17. http://plato.stanford.edu/entries/game-theory/#Uti
18. Y. Li and B. Bhanu. 2010. Utility-based camera assignment in a video network: a game-theoretic framework. *IEEE Sensors Journal*, 11.
19. G.R. Bradski. 1998. Computer vision face tracking for use in a perceptual user interface. *Intel Technology Journal*, 2nd quarter.
20. http://opencv.willowgarage.com/wiki/CvReference
21. M.J. Osborne and A. Rubinstein. 1994. *A Course in Game Theory*. Cambridge, MA: MIT Press.
22. F.Z. Qureshi and D. Terzopoulos. 2005. Surveillance camera scheduling: a virtual vision approach. *Third ACM International Workshop on Video Surveillance and Sensor Networks*.
23. J.R. Marden, P.H. Young, G. Arslan et al. 2009. Payoff-based dynamics for multi-player weakly acyclic games. *SIAM Journal on Control and Optimization*.
24. K. Morioka, S. Kovacs, J.H. Lee et al. 2008. Fuzzy-based camera selection for object tracking in a multi-camera system. *IEEE Conference on Human–System Interactions*.
25. Y. Li and B. Bhanu. 2009. Task-oriented camera assignment in a video network. *ICIP*.
26. B. Song, C. Soto, A.K. Roy-Chowdhury et al. 2008. Decentralized camera network control using game theory. *ACM/IEEE ICDSC Workshop on Embedded Middleware for Smart Camera and Visual Sensor Networks*.
27. Y. Li and B. Bhanu. 2010. On the performance of hand-off and tracking in a camera network. *ICPR*.
28. H. Nguyen, B. Bhanu, A. Patel et al. 2009. VideoWeb: design of a wireless camera network for real-time monitoring. *ACM/IEEE ICDSC*.
29. Y. Li and B. Bhanu. 2011. Fusion of multiple trackers in video networks. *Fifth IEEE/ACM International Conference on Distributed Smart Cameras*.

11 Cluster-Based Networking for MANETs Using Self-Organization

J. Guadalupe Olascuaga-Cabrera, Ernesto López-Mellado, Andres Mendez-Vazquez, and Félix Ramos-Corchado

CONTENTS

11.1 INTRODUCTION

Network systems are becoming increasingly complex due to their sizes, the heterogeneity of their underlying hardware, and the complex interactions among their elements. In recent years, many research efforts focused on virtual structures for ad hoc networks of wireless devices that have many important applications including disaster recovery, military operations, forest vigilance, and other uses. This type of network consists of a collection of nodes whose structure is dynamically formed "on the fly."

All the component nodes are functionally equal. Each node can act as a router and they do not need a centralized control. Therefore, virtual backbone construction is a widely used approach to ease implementation solutions to various problems such as routing and resource management strategies in ad hoc wireless networks. Current challenges of wireless networks include efficient networking and maintenance techniques aimed at saving energy and handling node mobility. Other important features desired for networking are segmentation recovery strategies and scalability, i.e., independence of the number of nodes.

Specifically, the hardware and energy limitations of wireless ad hoc networks require special management different from that used with wired networks. For example, they often are powered by batteries and have limited memory size and computational power. Therefore, memory usage and energy conservation are critical issues in these kinds of networks. In addition, the highly dynamic and autonomous environment of an ad hoc network calls for the design of algorithms that can operate in a distributed manner. Thus, the main characteristics of these new algorithms should include adaptability to network topology changes, robustness against failures, and the ability to operate in a localized and decentralized way.

Common features of current approaches include the high computational complexity of the strategies that requires high energy consumption and less ability to maintain network connectedness due to node mobility. To cope with these drawbacks, we propose drawing inspiration from biological systems that effectively adapt to constant and sudden changes in environments. Most of these systems consist of large numbers of dynamic, autonomous, and distributed entities that generate effective adaptive behaviors by using local policies such as self-organization. They cooperate with each other locally because global knowledge is not available within their neighborhood.

Several constraints regarding network connectivity and energy conservation must be considered. To deal with these constraints, self-organization strategies are effective for reducing the computational complexity and energy consumption of an entire ad hoc network.

11.2   NETWORKING APPROACHES

Pure flooding is simple and easy to implement. This approach provides high probability that each non-isolated node will receive the broadcast packet. The main disadvantage is that the technique consumes a large amount of bandwidth because of redundant retransmissions that diminish the lifetime of the network. Actually, not all the nodes retransmit a packet after receiving it.

Several algorithms have been proposed to optimize flooding. Examples are MPR (Ni et al. 1999; OK et al. 2006; Qayyum et al. 2000; Viennot 1998), CDS (Johnson and Carey 1979; Islam et al. 2009), and cluster-oriented strategies (Jamont et al. 2009; Dimokas et al. 2007; Heinzelman et al. 2000; Bajaber and Awan 2010; Younis et al. 2004; Mamei et al. 2006; Dressler 2008). We briefly review such approaches in this section.

11.2.1   MULTIPOINT RELAY (MPR)

The objective of MPR is to reduce the flooding of broadcast packets in a network by minimizing duplicate retransmissions locally (Mans and Shrestha 2004). Each node selects a subset of neighbors called MPRs to retransmit broadcast packets (see Figure 11.1). This allows neighbors that are not members of the MPR set to read a message without retransmitting it, thus preventing flooding of the network or "broadcast storm" (Ni et al. 1999; OK et al. 2006).

Of course, each node must select among its neighbors an MPR set that guarantees that all two-hop-away nodes will get the packets, i.e., all two-hop-away nodes must be neighbors of a node in the MPR set. Several polynomial time algorithms were proposed to select an MPR set with minimal cardinality. However, it was proven that the selection of a minimum size MPR set is an NP-hard problem (Qayyum et al. 2000; Viennot 1998).

11.2.2   CONNECTED DOMINATING SET (CDS)

CDS is used in a variety of applications, especially at the lower levels of a network protocol stack. Several CDS applications deal with the topology of wireless ad hoc networks (Nieberg and Hurink 2004). Given an undirected graph $G = (V, E)$, a subset

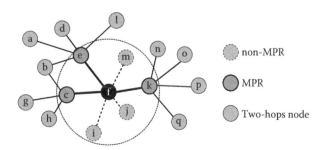

FIGURE 11.1   MPR strategy.

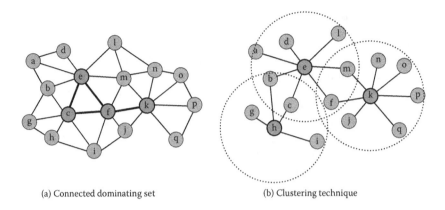

(a) Connected dominating set   (b) Clustering technique

FIGURE 11.2   CDS and clustering strategies.

$C \subseteq V$ is a CDS of G if, for each node $u \in V$, u is either in C or there exists a node $v \in C$ such that $uv \in E$ and the subgraph induced by C, i.e., $G(C)$, is connected. The nodes in the CDS are called *dominators* and the others are called *dominatees* (see Figure 11.2a).

Dominating sets play an important role in energy saving for individual sensors in wireless sensor networks. Obviously, the smaller the CDS, the fewer the retransmissions. However, a solution using dominating sets has one major drawback related to balanced energy distribution among the sensors: The sensors in a dominating set quickly consume their energy by sensing and transmitting data while the inactive sensors can save their energy. This causes disproportionate energy consumption among the sensors that is not desirable for applications in which network lifetime depends on the functioning of individual sensors. In addition, computing a minimum size CDS is also NP-hard (Johnson and Carey 1979).

The earlier problem was examined by Islam et al. (2009). The basis for this work was to find as many disjoint dominating sets as possible, use each one for a certain period, then replace it with another one, and so on. In general, it is expected that a dominating set will be as small as possible. Even though energy distribution is resolved, the environment used is short on mobility; if the nodes move around the environment, the strategy does not work and it will be necessary to execute the algorithm repeatedly to maintain the graph connection. Consequently, system complexity will increase. Aside from mobility problems, the nodes have homogeneous transmission power.

11.2.3   CLUSTER-BASED APPROACHES

Clustering techniques group the nodes into clusters; a cluster head ensures connectivity among the nodes within a cluster and with other clusters (see Figure 11.2b). Olariu et al. (2004) presented a light-weight clustering protocol for self-organization. Their technique creates a multi-hop, collision-free, adaptive communication infrastructure. The nodes wake up randomly, and the nodes that wake up earlier are designated cluster heads. After the nodes are organized into an adaptive multi-hop system, they perform tasks for saving energy. However, a large number of nodes is required to create an organization. This model does not deal with node mobility, role changes, or fault tolerance.

The strategies described above use diverse techniques, rules, and metrics. A common feature is that the transmission ranges are not variable. Using variable transmission ranges implies the creation of new rules for system organization to achieve energy saving (Thai et al. 2007). The nodes can transmit packets to every destination using only pertinent energy. Heinzelman et al. (2000) presented a micro-sensor network in which the base station is fixed and located far from the sensors. The authors examined two protocols: direct communication with the base station and minimum energy multi-hop routing using their sensor network and radio models.

By analyzing the advantages and disadvantages of conventional routing protocols using their sensor network model, they developed the low-energy adaptive clustering hierarchy (LEACH)—a cluster-based routing protocol that minimizes global energy usage by distributing the load to all the nodes at different times. However, it is necessary to know a priori the number of cluster heads in the network. All nodes in the network are homogeneous.

Younis and Fahmy (2004) presented a protocol called hybrid energy-efficient distributed (HEED) clustering. Its four primary goals are to: (1) increase network lifetime by distributing energy consumption, (2) terminate the clustering process within a constant number of iterations or steps, (3) minimize control overhead, and (4) produce well distributed cluster heads and compact clusters. This technique makes possible the identification of a set of cluster heads that covers the entire field, after which the nodes can communicate directly with their cluster heads via a single hop.

Kim et al. (2008) proposed a transmission power control algorithm for wireless sensor networks and call it the on-demand transmission power control (ODTPC) scheme. Link quality between a pair of nodes is measured after the sender and receiver exchange data ACK packets rather than measuring link quality to every neighbor in the initialization phase. No additional packet exchange is required to maintain good link quality and adjust transmission power. However, the authors were not creating a network structure.

Most of these studies assume that all the network information is known. Therefore, each strategy requires a lot of resources. These limitations led to the use of self-organization strategies to alleviate the energy consumption problem.

11.3 SELF-ORGANIZATION APPROACH

In wireless network applications, it is often necessary to keep the whole network connected while using a reduced number of resources. To interconnect a set of nodes, we propose a self-organization strategy that allows us to reduce network construction complexity and hence use less energy.

Self-organized systems consist of a large number of individual devices that interact and coordinate themselves to achieve tasks exceeding their capabilities as single individuals. The self-organizing phenomena can be found in nature, for example insect colonies and biological cells that organize their activities and achieve tasks with exceptional robustness despite harsh and dynamic environments (Mamei et al. 2006).

Natural systems capable of self-organization may exhibit emergent behavior involving a certain degree of order. These phenomena are well known in nature, for example, flocks of birds and the foraging behaviors of ants and schools of fish (Schmeck et al. 2010). The emergent property is defined as self-organized order based on the assumption that it can be observed as a pattern over time.

11.3.1 CONSTRUCTION OF NETWORK

A leader node is selected according to a weight that may correspond to a node's capability of performing additional duties. Leadership may be determined by assessing factors such as a node's residual energy, memory size, processing capabilities, or number of neighbors. Usually, the weights are computed locally in each node and may depend on the application for which the structure is used. In this case, weight is a function of the residual energy and the number of neighbors.

In this work, a node is modeled as an agent (the *node* and *agent* terms will be used interchangeably). Elements of each group play different roles (*leader, gateway, bridge,* and *member*). Each group consists of one agent acting as leader, zero or more agents playing the role of gateway and bridge, and one or more agents playing the role of member. The leader makes communication possible among members of its group or different groups. A gateway is responsible for communicating to members of different groups through group leaders. Bridge agents connect different segments of the network. Finally, members are connected to a single leader. Figure 11.3 shows the various agent roles. This model assumes that:

- Each agent has a unique identifier, for example, its IP address.
- Each agent discovers its one-hop neighbors.
- The agents can move, arrive, or leave the network.
- For purposes of energy conservation, each agent can adjust its transmission power according to its neighbors' roles.
- An agent can use overhearing to obtain important information for reducing message transmission.

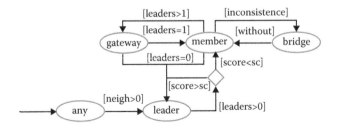

FIGURE 11.3 Different agent states.

- Each agent maintains a table of neighbors and a weight equal to the product of the number of neighbors N times the residual energy units e (*weight* = $N_u * e_u$).
- Agents do not know their geographical positions.
- Wireless agents are placed in a two-dimensional Euclidean plane. (This also works correctly if agents are located in three-dimensional spaces.)

11.4 PROPOSED STRATEGY

We propose a group-based distributed algorithm using a self-organization strategy that manages the assignments of four roles to be played by agents (Olascuaga-Cabrera et al. 2011). The procedure implementing the algorithm is executed by every agent in the network. To achieve this distributed processing, we decompose a desired global behavior into local rules to be followed by each entity. This requires defining and implementing local decision rules and entities that can act independently. The structure will yield scalable and robust systems that follow the desired global behavior.

When the agents wake up, they have no assigned roles. First, a role leader is assigned to all agents, then a conflict emerges and is solved by a leader election procedure. Some agents play the roles of members, and others act as gateways. To avoid network segmentation, we proposed a bridge agent role. When a member agent detects another member agent that belongs to another group, it will try to make a connection. Consequently, the agents will take on the bridge agent role. The bridge agent is connected to other bridge agents so they can prevent or fix possible segmentation. In this way, clusters are formed. The proposed algorithm has the following properties:

- It is an approximated decentralized algorithm.
- It operates in a localized way.
- It has low processing complexity.
- It allows network reconfiguration to cope with agent mobility.
- It ensures connectivity of the entire network.

11.4.1 GRAPH MODEL FOR WIRELESS NETWORK TOPOLOGY

A wireless ad hoc network is often represented by a graph in which an edge corresponds to a connection between two nodes. Graph theory presents many optimization problems that are important for dealing with efficient wireless communications strategies. We know that broadcasting based on a minimum spanning tree consumes energy within a constant factor of the optimum.

In this work, we studied a network as a series of disk graphs (Nieberg and Hurink 2004) in which all edges in the network are bidirectional. We assumed that nodes were deployed on a plane and modeled a wireless ad hoc network by an undirected graph $G = (V, E)$, where V is the set of nodes, each equipped with a radio for wireless communication. $E \leq V \times V$ is the set of possible interconnections between the nodes of each pair. For each edge $(u, v) \in E$, a weight $w(u, v)$ represents the cost to connect nodes u and v.

We assume that graph G has a *weight* function $w : E \rightarrow \mathbb{R}$. Then, using Prim's algorithm (Cormen et al. 2001), we find a minimum spanning tree T for G. Thus, we could use MST as a comparison for the proposed strategy.

We first provide a rule to calculate the *weight* function. N_u is the number of neighbors of node u, and e_u is the residual energy in a certain time period. If $(u, v) \notin E$, then $w(u, v) = \infty$, $c(u, v) = N_u * e_u$ (note that $c(u, v)$ is different from $c(v, u)$). Thus, the *weight* function is calculated:

$$w(u,v) = \frac{1}{c(u,v)+c(v,u)}$$

Definition 1. Let T be a subgraph of a graph G, such that it has been generated by MST (Prim's algorithm). A *backbone* of T, $MST_B^T T$ is $MST_B^T T = T \setminus S$, where $S = \{v \in T \mid deg_T (v) = 1\}$, and $deg_T (v)$ denotes the degree of vertex v.

Heterogeneous transmission ranges cause a graph to be directed. For this reason, we considered a solution induced only by symmetric edges, that is, two nodes are connected only if their separation is less than the minimum of their transmission ranges. An edge $(u, v) \in E$ if and only if $d(u, v) \le min\{r_u, r_v\}$ where $d(u, v)$ denotes the Euclidean distance between u and v and $r \in [r_{min}, r_{max}]$ is the transmission range.

11.4.2 NOTATION AND FUNCTIONS

The following list provides the necessary elements (notations and functions) to describe the self-organizing algorithm.

11.4.2.1 Events

$BR(u)$: bridge request
$BA(u)$: bridge acknowledge
$IR(u)$: check inconsistency

11.4.2.2 Sets

$N(u) = \{v \in V : (u, v) \in E\}$
$N_l(u) = \{v \in N(u) : Role(v) = Leader\}$
$N_g(u) = \{v \in N(u) : Role(v) = Gateway\}$

11.4.2.3 Cardinalities

$deg(u) = |N(u)|$: number of neighbors
$deg_L(u) = |N_l(u)|$: number of leader neighbors

11.4.2.4 Functions

$weight(u) = deg(u) * e_u$; where e_u is the residual energy of u
$id(u)$: returns agent's identification
$id_g(u)$: returns agent's group identifier
$Role(u)$: returns current role of agent u

11.4.2.5   Predicates

$C_0(u) = true\ if\ \exists\, v \in N(u) : Role(v) = Leader$
$C_1(u) = true\ if\ deg(u) > 0 \wedge \neg\, C_0(u)$
$C_2(u) = true\ if\ \exists\, v \in N(u) : Role(v) = Member \wedge id_g(u) \neq id_g(v)$
$C_3(u) = true\ if\ \exists\, v \in N(u) : Role(v) = Gateway \vee Role(v) = Bridge$
$C_4(u) = true\ if\ \exists\, v \in N_g(u) : N_l(u) = N_l(v) \wedge id(v) > id(u)$
$C_5(u) = true\ if\ \exists\, v \in N_g(u) : N_l(u) \subset N_l(v)$; where $\subset$ is the proper inclusion set operator

11.4.2.6   Rule-Based Role Assignment

The strategy that defines the updating of an agent's role is embedded in the following reduced set of rules.

$$R_1 : deg(u) = 0 \Rightarrow (Role_u \leftarrow Any)$$

$$R_1 : C_1(u) \Rightarrow (Role_u \leftarrow Leader)$$

$$R_3 : deg_L(u) = 1 \Rightarrow (Role_u \leftarrow Member)$$

$$R_4 : deg_L(u) > 1 \Rightarrow (Role_u \leftarrow Gateway)$$

$$R_5 : C_2(u) \wedge \neg\, C_3(u) \Rightarrow BR(v)$$

$$R_6 : BR(v) \wedge \neg\, C_3(u) \Rightarrow (Role_u \leftarrow Bridge)$$

$$R_7 : (C_0(u) \wedge Role(u) = Leader) \Rightarrow IR(v)$$

$$R_8 : IR(v) \wedge weight(v) > weight(u) \Rightarrow (Role_u \leftarrow Member)$$

$$R_9 : Role(u) = Member \wedge BR(v) \Rightarrow (BA(v) \wedge Role_u \leftarrow Bridge)$$

$$R_{10} : Role(u) = Gateway \wedge C_4(u) \Rightarrow agent_u\ turn\ off$$

$$R_{11} : Role(u) = Gateway \wedge C_5(u) \Rightarrow agent_u\ turn\ off$$

11.4.3   Procedure for Obtaining Virtual Backbone

At start-up, each agent broadcasts *hello* messages to determine its neighborhood. When an agent realizes the neighbor's table has not changed for a time σ, it will increase the broadcast transmission interval T by constant Δ to save energy.

When an agent has no neighbors, R_1 rule is applied and the agent acquires the *any* role. After an agent detects some neighbors, it has more than one rule to follow, depending on the neighbors' roles. If an agent's neighborhood has no leader, rule R_2 is applied and the agent acquires the role of leader. An agent becomes a member when R_3 or R_8 is applied, i.e., when it detects only one leader in its neighborhood or loses its leadership due to its low weight. Whenever an agent detects more than one leader, it changes its role from member to gateway through the application of rule R_4.

11.4.3.1 Leader Role

The most important role of an agent in a network is the leader. Leader agents are responsible for carrying out all the communication in the group. For this reason, they must have the highest weight because they cover most agents in the neighborhood and have the highest residual energy. For computing the agent's weight, we consider only the number of neighbors and the residual energy ($weight(u) = N_u * e_u$). This weight, however, may change based on the environment.

Initially, when an agent discovers at least one neighbor, the agent will seek a leader among its neighbors; if it does not find one, it becomes the leader (R_2). After that, when agents take on some of the available roles, a conflict may arise if two or more agents are leaders in the same group. When this situation happens (R_7), it is necessary to reach an agreement to select only one leader. To retain the leader role, an agent must have the highest weight in the group of conflicted agents (R_8).

11.4.3.2 Bridge Role

To carry out the bridge role, the agent must meet certain requirements. The choice is difficult because agents are aware only of their own neighborhoods. They do not perceive other agents beyond their one-hop neighbors. One objective of this algorithm is to maintain complete connectivity, but the network may be segmented.

A solution to solve the segmentation problem may be reorganization of segments in the network, i.e., every member that detects an inconsistency must inform its leader, after which the leader agent searches to identify the group involved in the segmentation problem. If the leader agent fails to find the identification, it will start the reorganization to alleviate the segmentation. However, every leader agent that detects an inconsistency will execute this procedure; this process consumes excessive time and energy. The previous solution wastes a great deal of energy due to excessive message usage and also fails to ensure good reorganization and overall connectivity in the affected subnetwork. To avoid these shortcomings, it is preferable to use bridge agents.

Every time a member agent u satisfies rule R_5, it generates an event that involves sending a bridge request message to v ($BR(v)$). After that, when a member agent u receives a bridge request from v, it becomes a bridge and returns a bridge acknowledge message to v (R_9). Now, when a member agent u receives an acknowledge message from v, it becomes a bridge agent, as shown in Figure 11.4d.

11.4.3.3 Gateway Role

The task of the gateway agent is communicating with leaders. In most cases, more than one gateway communicates with a single set of leaders. To alleviate the redundancy, a gateway agent will turn inactive if rule R_{10} or rule R_{11} is satisfied. Assume a gateway u turns inactive if the set of leaders in its neighborhood is equal to the set of leaders of one gateway neighbor v. At the same time, the id of u is lower than the id of v or the set of leaders of u is included within the set of leaders of one gateway neighbor.

11.4.4 ENERGY MANAGEMENT

Some member agents do not need maximum transmission power to communicate with their leaders. It is better for a member agent to adjust its transmission power to

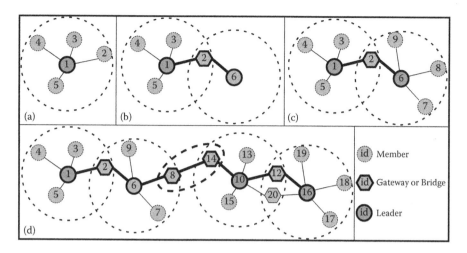

FIGURE 11.4 Clustering process.

reach its leader to ensure it uses only necessary energy (Figure 11.5). To calculate the optimal transmission power, we will use the following equation (Correia et al. 2007):

$$R_{TX\,min} = \frac{RX_{threshold} * P_{TX}}{P_{RX}}$$

where:

R_{TXmin} is the necessary transmission power to communicate with a leader agent.

$RX_{threshold}$ is the minimum signal strength to receive the packet. If one packet is received whose signal strength is stronger than $RX_{threshold}$, it is received correctly; otherwise it is discarded.

P_{TX} is the transmitted power.

P_{RX} is the received power.

Other factors influence communications in a network, for example, the background noise produced by signals naturally present in the environment. However, these factors are beyond the scope of this chapter.

11.4.5   SENSOR NETWORKS

Sensor networks for event detection have long periods of inactivity. They are also expected to be dense, and thus their redundancy can be exploited, allowing several agents to turn off their radios to conserve energy. However, connectivity must be maintained so that if an agent with an inactive radio senses something of importance, it can become active and successfully transmit the information. If member agents must sense and send information after long periods, it is not necessary to keep them awake. It is better to place them in sleep mode to save energy and increase network life.

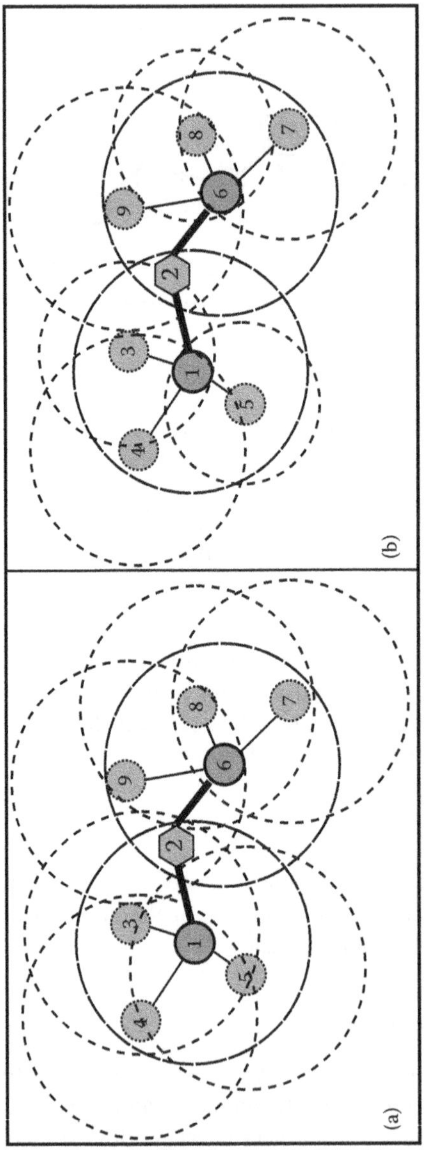

FIGURE 11.5 Reduced transmission power.

As we know, a leader agent needs more energy than other agents. That is why it is necessary to design a strategy to maintain its leadership as long as possible. After a leader agent dies, its neighborhood must be reconfigured. To delay the need for reconfiguration, a leader agent that has less energy than a threshold α will reduce its transmission range to 50% while it maintains at least two gateway or bridge agents to maintain communication. Whenever the energy residual of the leader agent is less than a threshold β, it will not maintain its role and will have to become a member agent.

We have focused on the minimum power topology problem in which the aim is to assign transmission power levels according to the agents' roles so that all agents are connected by bidirectional links and the total power consumption over a network is minimized. In short, the differences between the base algorithm (Jamont et al. 2009) and our strategies are:

- Member agents can adjust their transmission ranges.
- The leader holds the leadership for a longer time by varying its power transmission.
- Redundant gateways are turned off to save energy.
- All the agents vary their broadcast transmission intervals.
- The bridge role is designed to alleviate segmentation.

11.5   PERFORMANCE EVALUATION

This section shows that our algorithm behaves similarly to the MST. Although an MST is a sparsely connected subgraph, it is often not considered a good topology since nearby nodes in the original graph G may end up far away. If mobility is considered, we can argue that an algorithm should be both distributed and local. If the nodes move constantly, it is necessary to consider redundant links in the virtual backbone to maintain network connections. Redundant links allow reconfiguration of the virtual backbone using reduced energy.

11.5.1   EXPERIMENTS AND DISCUSSIONS

The behavior and performance of the proposed algorithm have been analyzed via simulation using NS-2 Version 2.33 (DARPA), an event-oriented simulator for network research.

11.5.1.1   Scenario

Consider a scenario in which a wireless sensor network must be organized. It consists of fifty mobile agents distributed within an environment of 50 × 50 meters (Figure 11.6a). For simplicity, let us assume that the agents have the same constraints (although this is not strictly required) and are turned on simultaneously. The initial configuration of the agents is:

- Network interface is 802.15.4.
- Initial energy of each node is 2 joules.

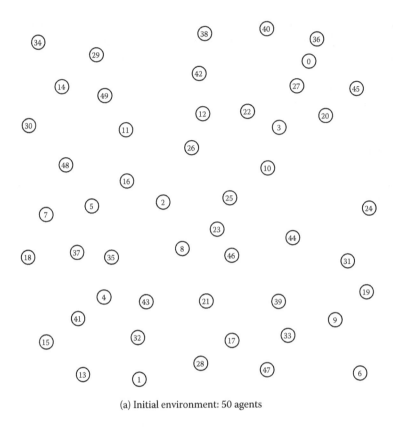

(a) Initial environment: 50 agents

FIGURE 11.6 Comparison of network connections.

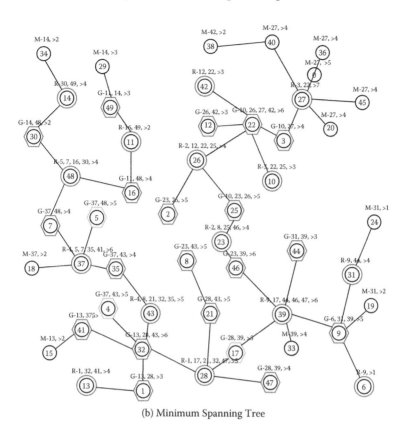

(b) Minimum Spanning Tree

FIGURE 11.6 (Continued)

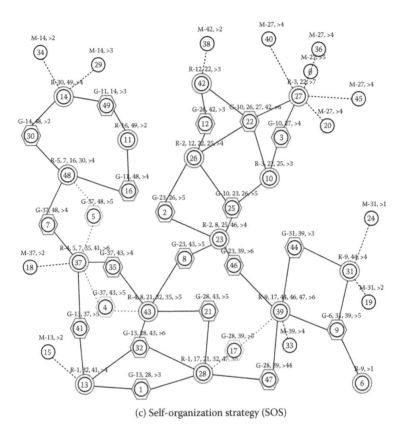

(c) Self-organization strategy (SOS)

FIGURE 11.6 (*Continued*)

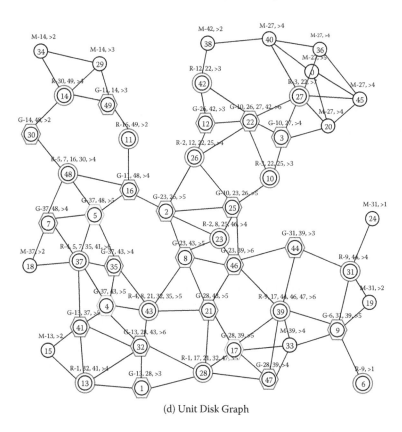

(d) Unit Disk Graph

FIGURE 11.6 (*Continued*)

- Tasks devoted to receiving messages, transmitting messages, and changing mode (active-sleep) will consume energy.
- Maximum transmission range is 10 meters.

Two versions of the proposed algorithm were tested. One considered static transmission power, and the other considered variable transmission power and variable transmission period.

11.5.1.2 Running NS-2

For comparison purposes a basic algorithm (Jamont et al. 2009) representative of related work was implemented. The algorithm has shown localized, distributed, and emergent behavior. The simulation in Figure 11.6c shows the backbone generated at instant $t = 14$ when the backbone connected the whole network. This behavior arose from using both versions of the algorithm. In environments where agents are turned on at the same time, a network can be created in segments because the agents interact only with their neighbors.

To correct this problem, the concept of bridge agents is introduced. However, it is not possible to predict how a network will behave during structuring. In fact, when an agent is removed, the restructuring may generate a different virtual backbone. In a situation in which agents are turned on simultaneously, a conflict arises when the nodes detect neighbors and decide to take on the role of leader. This will start a negotiation among the agents in conflict that will require high energy consumption. When the agents turn on asynchronously, the network is built gradually and yields a structure without problems.

In Figure 11.6c, fifty-six links connect the entire network without taking into account the gateway agents in sleeping state. When we construct a minimum spanning tree (Figure 11.6b) in the initial environment, we obtain a result similar to the proposed algorithm. The MST is created with forty-nine links among the agents. Both leader and gateway agents are in the virtual backbone.

The unit disk graph (UDG) $G(V, E)$ is typically used to model ad hoc wireless networks in this situation (Raei et al. 2008) if two agents are connected and their distance is within this fixed transmission range. Figure 11.6d shows the UDG in which all the agents are homogeneous, containing an edge (u, v) if and only if the unit disks around u and v intersect. The great difference between UDG and the simulation results is clear. The UDG has ninety-seven links and the proposed algorithm has forty-nine.

Based on Figure 11.6, our simulation results are close to those of the MST algorithm based on the number of links. In this way, we are close to optimum without using extra control messages. Instead of using a global strategy by having agents that retain knowledge of the whole environment; they are aware only of their neighborhoods.

11.5.1.3 Analyzing Energy Consumption

Total energy usage is defined as the total of the individual energy usage of each agent in the interval $[0, t]$, where t is the maximum simulation time. Network energy consumption levels for the entire network were compared using the two versions of

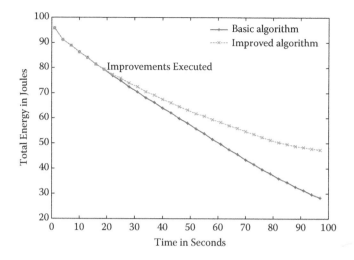

FIGURE 11.7   Improvement of total energy in network.

TABLE 11.1
Comparative Energy

Simulation	Range (m)	Agents	Dimensions	Initial Energy (Joules)	Energy Saving (%)
1	10 m	50	50 × 50	2	22
2	15 m	60	100 × 100	2	25
3	15 m	70	100 × 100	2	26
4	15 m	80	100 × 100	2	24
5	15 m	100	100 × 100	3	30

the protocol. The simulation time was $t = 100$ seconds. Clearly this time range was too short for real applications and was used only for comparison purposes. The energy behavior is shown in Figure 11.7. Note that at instant $\tau = 15$, the consumed energy was equal because $\tau = 15$ was the stabilization time of the network. This instant depends on the application.

After this time, the curves in the graphic differ. The lower curve corresponds to the energy consumption using the simplest algorithm. The upper curve shows the behavior of the energy consumption when the transmission power of the leader nodes is adjusted. In this curve, the energy savings at instant $\tau = 100$ is about 20%. Additionally, at time t, the connectivity is preserved.

This performance was demonstrated through several simulations using diverse scenarios. Table 11.1 summarizes some of these simulations. It is clear that energy saving is better with our strategy. The proposed algorithm saved 15 to 20% of energy in comparison to the basic algorithm.

TABLE 11.2

Comparative MST Performance

	Number of Links			Performance		
Simulation	MST	SOS	UDG	Inactive Links	Difference (Number of Links)	Proximity (%)
1	49	56	97	3	4	92
2	59	68	101	3	6	90
3	69	88	115	2	17	80
4	79	96	146	9	8	90
5	99	125	256	10	16	86

11.5.2 ANALYZING NETWORK TOPOLOGY

Minimizing the number of links in a computed virtual backbone can help to decrease control overhead since broadcasting for route discovery and topology update is restricted to a small subset of agents. Therefore, the broadcasting storm problem inherent to global flooding can be greatly decreased. In this way, a virtual backbone provides good resource conservation.

To evaluate the network formed by the proposed strategy, we compared it with the one obtained using the MST global procedure and then compared the number of links. Results for several scenarios are summarized in Table 11.2. Additionally, the links forming the backbone were compared. The number of links was close to the minimum required for obtaining connectivity in these simulations. The simulations allowed us to demonstrate that every agent in the MST_B^T set is included in the virtual backbone generated by the proposed strategy. Hence, the high resemblance between the structure in MST_B^T and the one generated by the self-organized strategy is evident.

11.5.3 SOLVING SEGMENTATION PROBLEM

To evaluate the performance of our algorithm under different parameters, we randomly deployed sixty agents in a fixed area of 100×100 m (Figure 11.8). The initial transmission range of each node was 15 m.

Figure 11.9a illustrates the simulation results without using bridge agents. We can see that the member agents 2 and 42 can see each other but have no connection. This kind of organization yields network segmentation and requires bridge agents.

Figures 11.9b through d compare the numbers of connections for three connected networks. Based on Figure 11.9b, the minimum spanning tree is made of fifty-nine links over the entire network. The simulation using bridge agents is shown in Figure 11.9c; the network is connected and uses sixty-eight links.

Specifically, the number of links obtained from the simulation was nine links more than the minimum spanning tree because communication was localized, i.e., the information used by the nodes was local. As expected, the results indicate that

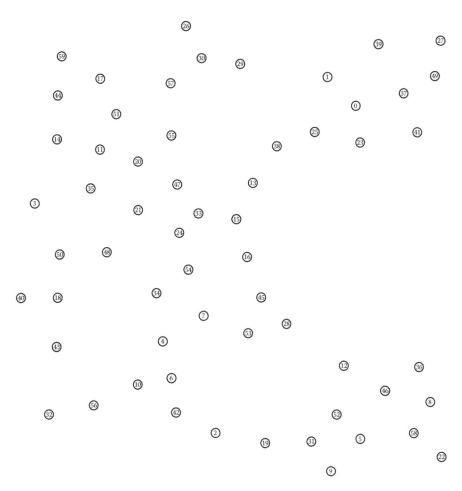

FIGURE 11.8   Initial environment: 60 agents.

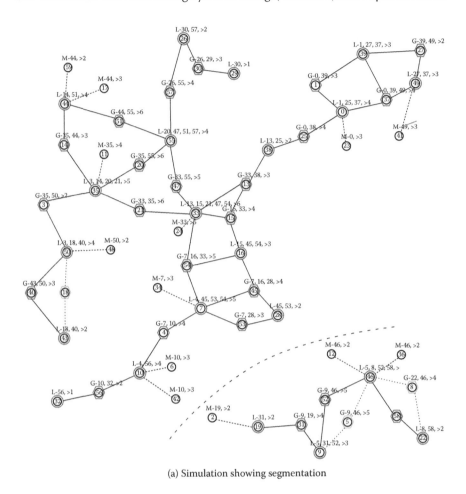

(a) Simulation showing segmentation

FIGURE 11.9 Simulation 2.

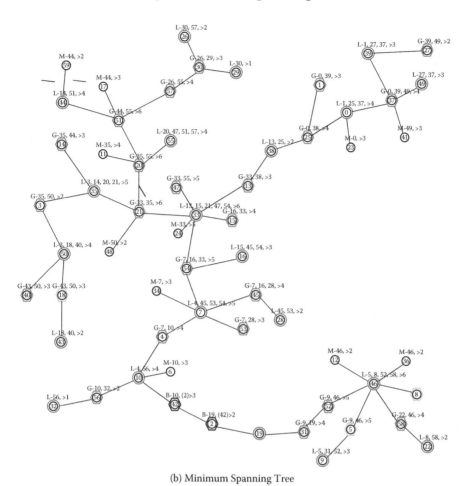

(b) Minimum Spanning Tree

FIGURE 11.9 (*Continued*)

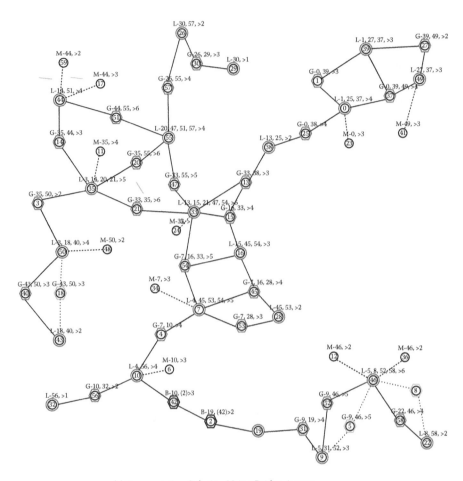

(c) Segmentation Solution Using Bridge Agents

FIGURE 11.9 (*Continued*)

(d) Unit disk graph

FIGURE 11.9 (*Continued*)

the simulation result approaches the minimum spanning tree. In addition, the unit disk graph shown in Figure 11.9d consists of 101 links—more than that yield in the simulation. These simulations reveal that results are consistent with our preliminary analysis. This means that we save energy without losing connectivity and also solve the segmentation problem of the basic approach.

11.6   CONCLUSION

Efficient use of energy is very important in ad hoc networks because most wireless devices are affected by this constraint. This chapter demonstrates that self-organization strategies improve energy savings during network formation and maintenance.

The proposed group-based algorithm handled variations in both transmission power and transmission periods. This allowed energy savings that increased the lifetime of the network. We also modeled the network as a graph $G = (V,E)$, and obtained the MST from G so we could compare it with the results of our algorithm. Simulations showed substantial improvements over the basic algorithm that does not use dynamic power transmission.

Our current research examines alternative strategies for reducing energy consumption and additional agent activities, namely routing, address allocation, and group-oriented bandwidth allocation.

REFERENCES

Bajaber, F. and Awan, I. 2010. Adaptive decentralized re-clustering protocol for wireless sensor networks. *Journal of Computer and System Sciences.*

Cormen, T.H., Leiserson, C.E., Rivest, R.L. et al. 2001. *Introduction to Algorithms.* Cambridge, MA: MIT Press.

Correia, L.H., Macedo, D.F., Santos, A.L. et al. 2007. Transmission power control techniques for wireless sensor networks. *Computer Networks,* 51, 4765–4779.

DARPA. n.d.. Retrieved from http://www.isi.edu/nsnam/ns/

Dimokas, N., Katsaros, D., and Manolopoulos, Y. 2007. Node clustering in wireless sensor networks by considering structural characteristics of the network graph. In *Proceedings of International Conference on Information Technology,* pp. 122–127.

Dressler, F. 2008. A study of self-organization mechanisms in ad hoc and sensor networks. *Elsevier Computer Communications,* 31, 3018–3029.

Heinzelman, W.R., Chandrakasan, A., and Balakrishnan, H. 2000. Energy-efficient communication protocol for wireless microsensor networks. In *Proceedings of 33rd Hawaii International Conference on System Sciences,* p. 8.

Islam, K., Akl, S.G., and Meijer, H. 2009. Maximizing the lifetimes of wireless sensor networks through domatic partition. In *Proceedings of 34th Annual IEEE Conference on Local Computer Networks,* pp. 436–442.

Jamont, J.P., Occello, M., and Lagreze, A. 2009. A multiagent approach to manage communication in wireless instrumentation systems. *Measurement,* 43, 489–503.

Johnson, D.S. and Carey, M.R. 1979. *Computers and Intractability: A Guide to the Theory of NP-Completeness.*

Kim, J., Chang, S., and Kwon, Y. 2008. ODTPC: on-demand transmission power control for wireless sensor networks. In *Proceedings of International Conference on Information Networking,* pp. 1–5.

Mamei, M., Menezes, R., Tolksdorf, R. et al. 2006. Case studies for self-organization in computer science. *Journal of Systems Architecture,* 52, 443–460.

Mans, B. and Shrestha, N. 2004. Performance evaluation of approximation algorithms for multipoint relay selection. In *Proceedings of Third Annual Mediterranean Ad Hoc Networking Workshop,* pp. 480–491.

Ni, S.Y., Tseng, Y.C., Chen, Y.S. et al. 1999. The broadcast storm problem in a mobile ad hoc network. In *Proceedings of Fifth Annual ACM/IEEE International Conference on Mobile Computing and Networking,* pp. 151–162.

Nieberg, T. and Hurink, J. 2004. Wireless communication graphs. In *Proceedings of Intelligent Sensors, Sensor Networks and Information Processing Conference,* pp. 367–372.

O.K., T., N., W., J.S., P. et al. 2006. On the broadcast storm problem in ad hoc wireless networks. In *Proceedings of Third International Conference on Broadband Communications, Networks and Systems,* pp. 1–11.

Olariu, S., Xu, Q., and Zomaya, A.Y. 2004. An energy-efficient self-organization protocol for wireless sensor networks. In *Proceedings of Intelligent Sensors, Sensor Networks, and Information Processing Conference,* pp. 55–60.

Olascuaga-Cabrera, J., López-Mellado, E., Mendez-Vazquez, A. et al. 2011. A self-organization algorithm for robust networking of wireless devices. *IEEE Sensors Journal*, 11, 771–780.

Qayyum, A., Viennot, and Laouiti, A. 2000. Multipoint relaying: an efficient technique for flooding in mobile wireless networks. (RR-3898).

Raei, H., Sarram, M., Salimi, B. et al. 2008. Energy-aware distributed algorithm for virtual backbone in wireless sensor networks. In *Proceedings of International Conference on Innovations in Information Technology*, pp. 435–439.

Schmeck, H., Muller-Schloer, C., Cakar, E. et al. 2010. Adaptivity and self-organization in organic computing systems. *ACM Transactions on Autonomous and Adaptive Systems*, 5, 10:1–10.32.

Thai, M.T., Wang, F., Liu, D. et al. 2007. Connected dominating sets in wireless networks with different transmission ranges. *IEEE Transactions on Mobile Computing*, 6, 721–730.

Viennot, L. 1998. Complexity results on election of multipoint relays in wireless networks. *Wireless Networks*, Report RR-3584, INRIA.

Younis, O., Member, S., and Fahmy, S. 2004. HEED: a hybrid energy-efficient distributed clustering approach for ad hoc sensor networks. *IEEE Transactions on Mobile Computing*, 3, 366–379.

12 Ensemble-Based Approach for Targeting Mobile Sensor Networks: Algorithm and Sensitivity Analysis

Han-Lim Choi and Jonathan P. How

CONTENTS

12.1 INTRODUCTION

This chapter addresses targeting of mobile sensor networks to extract maximum information from an environment in the context of numerical weather prediction. Accurate prediction of a weather system typically requires extensive in situ observations distributed over a large spatial and temporal domain because of huge dimensionality, nonlinearity, and uncertainties in a system. Without sufficient measurements, the difference between estimated and true state of a system can grow rapidly, leading to arbitrary and inaccurate predictions.

Complete sensor coverage in such systems is impractical because of excessive cost, time, and effort. Also, domain characteristics (such as geographic features) often limit how sensors may be deployed. For example, it is far more difficult to deploy permanent weather sensors over open ocean areas than on land.

One popular approach for dealing with this difficulty is to expand the weather observation network by augmenting fixed networks with adaptive observations via redeployable (possibly mobile) sensors. In particular, based on the enhanced endurance and autonomy of unmanned aircraft systems (UASs), similar devices have become increasingly attractive for redeployable weather sensing platforms [1,2]. The aerospace robotics and control literature discusses a number of new developments of the planning algorithms. Examples are designs for path planning algorithms to avoid undesirable weather conditions such as icing [3,4] and guidance algorithms for networks of severe storm penetrators for volumetric atmospheric sensing [5,6]. The Tempest UAS succeeded in its first encounter with a tornadic supercell [7].

However, these results have not considered the vital decisions of where and when to utilize redeployable sensors to improve the quality of weather forecasts. This decision making process was addressed in the numerical weather prediction literature [8–16], primarily discussing concurrent measurement platforms (e.g., two-crewed aircraft) that were weakly coordinated.

Our recent work deals with these types of decisions for UAS sensor networks [17–21]. Our intention was to improve the efficiency of such decisions and thus facilitate more adaptive and coordinated allocation of mobile sensing resources in large-scale natural domains. This chapter presents the algorithm and sensitivity analysis of an ensemble-based targeting method whose main results were reported [19,21].

This work proposes an efficient way to design targeted paths for mobile sensor networks when the goal is to reduce the uncertainty in a forecast arising from some verification region at some verification time. Mutual information is used as a measure of uncertainty reduction and it is computed from the ensemble approximation [9–11,15,22,23] of covariance matrices. The method is built on the backward formulation proposed in our earlier work [20] on efficient quantification of mutual information in a combinatorial targeting process.

The methodology in this article extends the backward formulation by providing mechanisms to effectively handle the constrained mobility of UAS sensor platforms. A reachable search space is calculated based on the minimum and maximum speeds of the vehicles; this space constitutes the minimal set for which the associated covariance (or ensemble) updates must be performed.

The search process goes through the action space (instead of the measurement space) to exclude infeasible solutions. A cut-off process is then proposed along with a simple cost-to-go heuristic that approximates mutual information using some sparsity assumption. These three mechanisms offer improvements in computational efficiency of a constrained targeting process.

This chapter also identifies potential issues in implementation of the ensemble-based targeting method to real-scale weather forecasting. Due to the computational expense of integrating a large-scale nonlinear system and storing large ensemble data sets, the ensemble size that can be used for adaptive sampling in real weather prediction systems is very limited [24–26]. Despite this limitation, no research focuses on the sensitivity of a targeting solution to ensemble size.

As an essential step toward implementation of the presented information-theoretic ensemble targeting algorithm to realistic cases, this work performs sensitivity analysis of ensemble-based targeting based on limited ensemble size. Statistical analysis of the entropy estimation informs a predictor of the impact that limitation of ensemble size may exert on the optimality of the solution to a targeting problem.

It should be noted that this chapter is based on two recent articles of the authors [19,21]. Most of the text and figures in Sections 12.2 through 12.4 are adapted from Reference [19]. Preliminary work cited in Section 12.6 was presented in Reference [21].

12.2 BACKGROUND

12.2.1 ENTROPY AND MUTUAL INFORMATION

The entropy that represents the amount of information hidden in a random variable can be expressed specifically via a covariance matrix as:

$$\mathcal{H}(A) = \frac{1}{2}\log\det(\mathrm{Cov}(A)) + \frac{|A|}{2}\log(2\pi e)$$

for a Gaussian random vector A. The covariance matrix $\mathrm{Cov}(A) \triangleq \mathbb{E}\left[(A - \mathbb{E}[A])(A - \mathbb{E}[A])^T\right]$ where the superscript T denotes the transpose of a matrix and another notation $\mathrm{Cov}(A,B)$ will be used to represent $\mathrm{Cov}(A,B) \triangleq \mathbb{E}\left[(A - \mathbb{E}[A])(B - \mathbb{E}[B])^T\right]$; $|A|$ denotes the cardinality of A, and e is the base of a natural logarithm.

The mutual information represents the amount of information contained in one random variable (A_2) about the other random variable (A_1), and can be interpreted as the entropy reduction of A_1 by conditioning on A_2:

$$\mathcal{I}(A_1; A_2) = \mathcal{H}(A_1) - \mathcal{H}(A_1 \mid A_2) \tag{12.1}$$

where the last term represents the conditional entropy of A_1 conditioned on A_2. It is important to note that the mutual information is commutative [27]:

$$\mathcal{I}(A_1; A_2) = \mathcal{I}(A_2; A_1) \tag{12.2}$$

In other words, the entropy reduction of A_1 by knowledge of A_2 is the same as the entropy reduction of A_2 by knowledge of A_1. Because of this symmetry, the mutual information can also be interpreted as the degree of dependency between two random variables.

12.2.2 ENSEMBLE SQUARE-ROOT FILTER

The weather variables are tracked by an ensemble forecast system, specifically the sequential ensemble square-root filter (EnSRF) [22]. Ensemble-based forecasts better represent the nonlinear features of a weather system and mitigate the computational burdens of linearizing the nonlinear dynamics and keeping track of a large covariance matrix involved in an extended Kalman filter [9,22,23]. EnSRF carries the ensemble matrix:

$$\mathbf{X} = \left[\mathbf{x}_1, \dots, \mathbf{x}_{L_E} \right] \in \mathbb{R}^{L_S \times L_E}$$

where $\mathbf{x}_i$ for $i \in \{1, \dots, L_E\}$ is the i-th sample state representation; L_S and L_E denote the number of state variables and number of ensemble members, respectively. With the ensemble matrix, the state estimation and estimation error covariance are represented by the ensemble mean $\bar{\mathbf{x}}$ and the perturbation ensemble matrix $\widetilde{\mathbf{X}}$, written as:

$$\bar{\mathbf{X}} \triangleq \frac{1}{L_E} \sum_{i=1}^{L_E} \mathbf{x}_i, \; \widetilde{\mathbf{X}} \triangleq \eta(\mathbf{X} - \bar{\mathbf{x}} \otimes \mathbf{1}_{L_E}^T) \tag{12.3}$$

where $\otimes$ denotes the Kronecker product and $\mathbf{1}_{L_E}$ is the L_E-dimensional column vector whose entries are all ones. η is an inflation factor chosen to be large enough to avoid underestimation of the covariance and small enough to avoid divergence of filtering [22]; $\eta = 1.03$ is used in this chapter. Using the perturbation ensemble matrix, the estimation error covariance is approximated as:

$$\text{Cov}(\mathbf{x}) \approx \tilde{\mathbf{x}}\widetilde{\mathbf{X}}^T/(L_E - 1). \tag{12.4}$$

The conditional covariance is calculated as the following sequential measurement update procedure. The sequential framework is devised for efficient implementation [22]. Let the j-th observation be the measurement of the i-th state variable, and R_i be the associated sensing noise variance. The ensemble update for the j-th observation is given by:

$$\widetilde{\mathbf{X}}^{j+1} = \widetilde{\mathbf{X}}^j - \gamma_1\gamma_2\widetilde{\mathbf{X}}^j\xi_i^T\xi_i/(L_E - 1) \tag{12.5}$$

with $\gamma_1 = (1 + \sqrt{\gamma_2 R_i})^{-1}, \gamma_2 = (p_{ii} + R_i)^{-1}, \xi_i$ is the i-th row of $\widetilde{\mathbf{X}}^j$ and $p_{ii} = \xi_i\xi_i^T/(L_E - 1)$. γ_1 is the factor for compensating the mismatch between the serial update and the batch update, and $\gamma_2\widetilde{\mathbf{X}}^j\xi_i^T$ is equivalent to the Kalman gain.

12.3 PROBLEM FORMULATION

12.3.1 COORDINATED SENSOR TARGETING

The coordinated sensor network targeting problem considers gridded space–time of finite dimension (Figure 12.1). The location of a vehicle can be described as a positive integer by having the index set defined such that $r = s + \Delta t \times n_G$ where r and s represent two points that are spatially identical but temporally apart by Δt time steps, which is positive when representing posteriority; n_G denotes the spatial dimension of the grid representation.

In this chapter, the search space–time $S \subset \mathbb{Z}$ is defined as the set of location indices at which a sensor platform may be located. Without loss of generality, it is assumed that a grid point s is associated with a single state variable X_s that can be directly measured; the measurement at this location is denoted as Z_s and $Z_s = X_s + W_s$ where W_s is the white Gaussian noise with variance R_s that is independent of sensing noise at other locations and of any state variable.

Likewise, V represents the set of verification variables at verification time t_V. We denote the location index of the i-th sensor platforms at time t_k as $S_i[t_k] \in \mathbb{Z}$ and the number of sensor platforms as n_s. The set of locations of all the sensor platforms at time instance t_k is denoted as $\mathbf{s}[t_k] \in \mathbb{Z}^{n_s}$.

The transition of a sensor platform's location at one time step to the next time step can be written as $s_i[t_{k+1}] = s_i[t_k] + u_i[t_k]$ where $u_i[t_k] \in \mathcal{U} \subset \mathbb{Z}$ is the control action taken by the sensor platform i at time step t_k.

The set $\mathcal{U}$ defines all the possible control options that can be represented as positive integer values in the grid space–time. Since a vehicle has limited capability of motion, $\{\mathcal{U}\}$ should be a finite set. Also, the control vector is defined as:

$$\mathbf{u}[t_k] \triangleq [u_1[t_k], u_2[t_k], \ldots, u_{ns}[t_k]]^T.$$

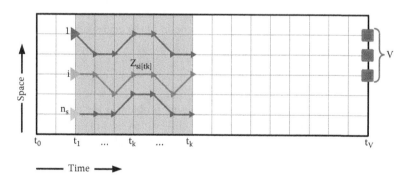

FIGURE 12.1 Coordinated targeting of mobile sensors in space–time grid. The adaptive observation targeting problem decides the best additional sensor network over the time window $[t_1, t_K]$ given the routine network that makes observations over the same time window. The routine observations between $[t_K, t_V]$ are not of practical concern because the ultimate interest is in the quality of a forecast broadcast at t_K based on the actual measurement taken up to time t_K (M = size of verification region, n_s = number of sensor platforms, K = number of waypoints per sensor). (*Source:* IEEE© 2011.)

The goal of targeting is to find the optimal control sequences $\mathbf{u}[t_k]$, $k \in \{1, ..., K-1\}$ (or equivalently optimal waypoint sequences $s[t_k]$, $\forall k \in \{1, ..., K\}$) to maximize the information gain to best predict the verification variables V. In this work, mutual information is adopted to quantify the amount of information that can be extracted from measuring environmental variables.

The mathematical formulation of the targeting problem can be written as the following constrained optimization:

$$\max_{\mathbf{u}[t_1], ..., \mathbf{u}[t_{K-1}]} \mathcal{I}(V; Z_{s[t_1]}, ..., Z_{s[t_K]}) \tag{12.6}$$

subject to

$$s[t_{k+1}] = s[t_k] + \mathbf{u}[t_k], \forall k \in \{1, 2, ..., K-1\}. \tag{12.7}$$

$$s[t_1] = s_o = \text{given}, \tag{12.8}$$

and

$$s_i[t_k] \in S, \forall i \in \{1, 2, ..., n_s\}, \forall k \in \{1, 2, ..., K\}. \tag{12.9}$$

The objective function of Equation (12.6) is the mutual information between the verification variables V and the measurement sequence $\cup_{k=1}^{K} Z_{s[t_k]}$ taken by the UAV sensors from time t_1 through t_K. The condition in Equation (12.7) describes the vehicle dynamics when initial (t_1) locations of vehicles are assumed to be given as in Equation (12.8).

With regard to the control options of the sensor platforms, this work focuses on the limited mobility of a sensor platform; thus, a legitimate control action leads the vehicle to one of its neighboring locations at the next time step. The condition in Equation (12.9) means that vehicles should remain in the grid space by their control actions—we can call it the *admissibility* condition.

12.3.2 ENSEMBLE-BASED FORMULATION

The prior knowledge in the observation targeting problem is the analysis ensemble at t_0. Since the design of the supplementary observation network is conditioned on the presence of a routine network, the covariance field needed for sensor selection is the conditional covariance field conditioned on the impacts of future routine observations. Processing the observations distributed over time amounts to an EnSRF ensemble update for the augmented forecast ensemble defined as:

$$\mathbf{X}_{aug}^f = \left[(\mathbf{X}_{t_1}^f)^T, ..., (\mathbf{X}_{t_K}^f)^T, (\mathbf{X}_{t_V}^f)^T \right]^T \in \mathbb{R}^{(K=1)L_S \times L_E},$$

where $\mathbf{X}_{t_k}^f = \mathbf{X}_{t_0}^a + \int_{t_0}^{t_k} \dot{\mathbf{X}}(t)dt$. $\mathbf{X}_{t_0}^a \in \mathbb{R}^{L_S \times L_E}$ is the analysis ensemble at t_0. For computational efficiency, the impact of future routine observations is processed in advance and the outcome provides the prior information for the selection process. The routine observation matrix for the augmented system is expressed as:

$$\mathbf{H}_{aug}^r = \left[\mathbf{H}^r \otimes I_K \ \mathbf{0}_{Kn_r \times L_S} \right], \quad \mathbf{H}^r \in \mathbb{R}^{n_r \times L_S}$$

where n_r is the number of routine measurements at each time step. Since only the covariance information is needed (and available) for targeting, incorporation of the future routine networks involves only the perturbation ensemble update:

$$\widetilde{\mathbf{X}}_{aug}^a = (I - \mathbf{K}_{aug}^r \mathbf{H}_{aug}^r) \widetilde{\mathbf{X}}_{aug}^f$$

without updating the ensemble mean. In the sequential EnSRF scheme, this process can be performed one observation at a time using Equation (12.5). The later sections of this chapter assume that $\widetilde{\mathbf{X}}_{aug}^a$ was computed before posing the targeting problem in Equation (12.6).

12.4 TARGETING ALGORITHM WITH MOBILITY CONSTRAINTS

This section presents an algorithm to solve the mobile sensor targeting problem formulated in Section 12.3. The main features of the proposed algorithm are (1) the backward selection formulation in computing a mutual information value for the solution candidates, (2) the reachable search space and action space search to effectively incorporate vehicle mobility constraints, (3) a cut-off heuristic for reducing the number of solution candidates for which objective values are actually calculated. Details are presented along with a summary of the algorithm.

12.4.1 EFFICIENT SENSOR SELECTION FRAMEWORK

The quantification of the mutual information for each measurement sequence is based on the *backward* formulation [20] we developed. The concept was inspired by the commutativity of mutual information. While a straightforward formulation considers how much uncertainty in the verification variables V would be reduced by a potential measurement sequence, the backward approach looks at the reduction of entropy in the measurement variables instead of the verification variables. The mutual information is calculated as:

$$\mathcal{I}(V; Z_{s[t_1:t_K]}) = \mathcal{H}(Z_{s[t_1:t_K]}) - \mathcal{H}(Z_{s[t_1:t_K]} \mid V)$$

$$= \frac{1}{2} \log \det \left(\mathrm{Cov}(X_{s[t_1:t_K]}) + R_{s[t_1:t_K]} \right) - \frac{1}{2} \log \det \left(\mathrm{Cov}(X_{s[t_1:t_K]} \mid V) + R_{s[t_1:t_K]} \right).$$

$$(12.10)$$

Since the conditional covariance matrix of the state variables in the search space $\mathrm{Cov}(X_{s[t_1:t_K]} \mid V)$ can be extracted from the conditional covariance over the entire search space $\mathrm{Cov}(X_{s[t_1:t_K]} \mid V)$ without further computation, this alternative formulation reduces the number of covariance updates to one. This computational benefit is significant when the description of the environmental variables is in ensemble form. The ensemble-based backward selection is expressed as:

$$\mathcal{I}(V; Z_{s[t_1 : t_K]}) = \frac{1}{2} \log \det \left(\frac{1}{L_E - 1} \tilde{\mathbf{x}}_{X_{s[t_1 : t_K]}} \tilde{\mathbf{x}}^T_{X_{s[t_1 : t_K]}} + R_{s[t_1 : t_K]} \right)$$
$$- \frac{1}{2} \log \det \left(\frac{1}{L_E - 1} \tilde{\mathbf{x}}_{X_{s[t_1 : t_K]}} \mid V \; \tilde{\mathbf{x}}^T_{X_{s[t_1 : t_K]}} \mid V + R_{X_{s[t_1 : t_K]}} \right)$$

(12.11)

and we have proved that the relative computational efficient is obtained as:

$$\frac{T_{\text{fwd,ens}}}{T_{\text{bwd,ens}}} > \frac{9}{2} L_E / n$$

where $T_{\text{fwd,ens}}$ and $T_{\text{bwd,ens}}$ denote computation times for the ensemble-based forward and backward selection process, respectively, and n is the number of sensing points that are simultaneously selected. Note that L_E tends to be substantially greater than n as the description of a large-scale system requires sufficient points (although limited by computational resource availability as discussed in Section 12.6) and n is a design parameter that can be tuned for computational tractability as discussed in Section 12.4.5.

12.4.2 REACHABLE SEARCH SPACE

In the backward selection framework, a covariance update is needed to compute the posterior covariance over the search space by fictitious measurements at V. Because the motions of sensor platforms are limited, not all the points in S can be reached by the sensor platforms; therefore, a set smaller than S effectively works as the search space over which the posterior covariance is calculated. This reduced search space should be as small as possible, but must be sufficiently exhaustive to consider all the admissible and reachable candidates.

To find the smallest possible search space, first define the one-step reachable set that represents the set of all possible location vectors that the team of sensor platforms at location $s[t_k]$ at time t_k can reach at the next time step t_{k+1} by applying a legitimate control vector:

$$\mathcal{R}(s[t_k]) = \left\{ \mathbf{r} \subset \mathcal{S} \mid \mathbf{r} = s[t_k] + \mathbf{u}[t_k]; s[t_k] \subset \mathcal{S}, \mathbf{u}[t_k] \subset \mathcal{U}^{n_s} \right\}.$$

The *reachable search space* denoted as $\mathcal{S}_Z$ and consists of all the points in S that can be visited by some sensor platform by some control actions defined as:

$$\mathcal{S}_Z = \bigcup_{k=1}^{K-1} \mathcal{R}^k(\mathbf{s}_0),$$

(12.12)

where $\mathbf{s}_0$ is the initial vehicle location at time t_1 as defined in Equation (12.8) and $\mathcal{R}^{k+1}(\mathbf{s}_0) = \mathcal{R}\left(\mathcal{R}^k(\mathbf{s}_0) \right)$, $\forall k \in \{1, 2, \ldots, K-1\}$. This $\mathcal{S}_Z$ is minimal in the sense that every element in $\mathcal{S}_Z$ will be cited at least once in computing the mutual information for a feasible solution candidate. Since the right side of Equation (12.12) is a union of disjoint sets, the cardinality of $\mathcal{S}_Z$ becomes:

$$|\mathcal{S}_Z| = \sum_{k=1}^{K-1} |\mathcal{R}^k(s_o)| \leq \sum_{k=1}^{K-1} |\mathcal{U}|^{n_s k} \sim \mathcal{O}\left(|\mathcal{U}|^{n_s(K-1)}\right).$$

Note that the size of $\mathcal{S}_Z$ is exponential in n_s and $K - 1$; thus, the optimal targeting problem with many sensor platforms for a long time horizon needs a larger $\mathcal{S}_Z$. This exponential dependency requires decomposition schemes that will be discussed in Section 12.4.5.

12.4.2.1 Minimum and Maximum Speed Case

As an example of the vehicle mobility constraints, this work considers the minimum and maximum speeds of a sensor platform.

$$v_{min} \leq \mathcal{D}_M(s_i[t_k], s_i[t_{k+1}]) \leq v_{max}. \tag{12.13}$$

$\mathcal{D}_M$ denotes the Manhattan distance defined as

$$\mathcal{D}_M(s,r) = \sum_{j=1}^{d} |s^j - r^j|$$

where d denotes the dimension of spatial coordinates; s^j and r^j denote the j-th spatial coordinate values of the grid points s and r in S, respectively. Figure 12.2 illustrates the reachable zones by a single agent from location s in a two-dimensional grid space in case $v_{min} = 1$ and $v_{max} = 2$. Note that:

$$\mathcal{R}^k(s_o) = \{r \in \mathcal{S}[t_{k+1}] \mid \mathcal{D}_M(s,r) \leq k \times v_{max}\}$$

for all $k \in \{2,\ldots,K-1\}$ where $\mathcal{S}[t_{k+1}] \subset S$ denotes the set of grid points whose time index is t_{k+1}. Thus, the size of $\mathcal{S}_Z$ for a single sensor in a two-dimensional space becomes:

$$|\mathcal{S}_Z|_{single} = \sum_{k=2}^{K-1} \left(1 + 4\sum_{j=1}^{kv_{max}} j\right) + 4\sum_{j=v_{min}}^{v_{max}} j.$$

If each sensor platform has identical moving capability, the reachable zone by a team of sensor platforms is nothing more than the union of each agent's reachable zone:

$$\mathcal{S}_Z(s_o) = \bigcup_{i=1}^{n_s} \mathcal{S}_Z(s_i[t_1]),$$

which is smaller than or equal to $|\mathcal{S}_Z| \leq n_s \times |\mathcal{S}_Z|_{single}$.

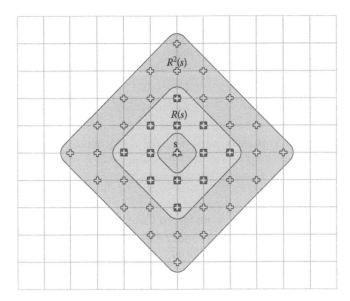

▲ Initial vehicle location: s

■ Points reachable by one movement

✚ Points reachable by two movements

FIGURE 12.2 Reachable zones by single agent in two-dimensional grid space. (*Source:* IEEE© 2011.)

12.4.3 Action Space Search

With a fixed initial location vector $\mathbf{s}_o$, the targeting problem finds the optimal $\mathbf{s}[t_2 : t_K]$ consisting of $n_s(K - 1)$ grid points. The reachable search space $\mathcal{S}_Z$ defines the set over which the covariance update must be done. Every point in $\mathcal{S}_Z$ can be reached by some control actions of some sensor platform. However, this does not mean that *every* set $\mathbf{s} \subset \mathcal{S}_Z$ with $|\mathbf{s}| = n_s(K - 1)$ comprises some feasible waypoint sequences. Thus, an n-choose-k-type search over $\mathcal{S}_Z$ is not an efficient method when vehicle motions are constrained.

For this reason, this work suggests a search over the *action space*. The search steps are (1) pick a feasible control sequence $\mathbf{u}[t_1 : t_K]$; (2) find a corresponding location vector sequence $\mathbf{s}[t_1 : t_K]$; and (3) evaluate $\mathcal{I}(Z_{\mathbf{s}[t_1 : t_K]}; V)$. When using the action space search, the total number of solution candidates to consider becomes:

$$\text{NumberCandidates} = |\mathcal{U}|^{n_s(K-1)}.$$

This number is typically much smaller than $\begin{pmatrix} |\mathcal{S}_Z| \\ n_s(K-1) \end{pmatrix}$ for the n-choose-k-type search, which could be $\mathcal{O}(|\mathcal{U}|^{n_s^2(K-1)^2})$ in the worst case.

12.4.4 CUT-OFF HEURISTICS

Although the backward approach described in Section 12.4.1 does not involve a combinatorial number of covariance updates, it still requires the computation of the determinant of:

$$\text{Cov}(Z_{s[t_2:t_K]} \mid Zs_0), \quad \text{Cov}(Z_{s[t_2:t_K]} \mid V, Zs_0), \tag{12.14}$$

for a total of $|\mathcal{U}|^{n_s(K-1)}$ times. For instance, with the mobility constraint in Equation (12.13),

$$|\mathcal{U}|^{n_s(K-1)} = \left[4 \sum_{j=v_{min}}^{v_{max}} j\right]^{n_s(K-1)} = \left[2(v_{max} + v_{min})(v_{max} - v_{min} + 1)\right]^{n_s(K-1)},$$

for a two-dimensional problem. Therefore, if v_{max} and/or $n_s(K-1)$ becomes large, the number of covariance matrices in Equation (12.10) to be considered in the decision rapidly grows. Moreover, as $n_s(K-1)$ becomes larger, the unit time for computing the determinant of one covariance matrix also increases proportionally to $n^3(K-1)^3$ [28].

To address a large-scale problem, it is necessary to reduce the number of solution candidates whose mutual information values (determinants of prior and posterior covariance matrices) are actually calculated. For this reason, this chapter proposes a cutoff heuristic that provides an indication of which measurement choice would render a high information reward. For this purpose, an approximation of

$$\tilde{\mathcal{I}}_{s[t_2:t_K]} \triangleq \tilde{\mathcal{H}}^-_{s[t_2:t_K]} - \tilde{\mathcal{H}}^+_{s[t_2:t_K]} \tag{12.15}$$

as an indication of whether candidate $s[t_2: t_K]$ is worth considering in the mutual information computation where the two terms on the right side represents the prior and the posterior entropies of $Z_{s[t_2:t_K]}$ obtained under the assumption that the associated covariance matrices are diagonal.

$$\tilde{\mathcal{H}}^-_{s[t_2:t_K]} = \frac{1}{2} \sum_{i=1}^{n_s(K-1)} \log\left[P^-_{s[t_2:t_K]}(i,i) + R_{s[t_2:t_K]}(i,i)\right]$$

$$\tilde{\mathcal{H}}^+_{s[t_2:t_K]} = \frac{1}{2} \sum_{i=1}^{n_s(K-1)} \log\left[P^+_{s[t_2:t_K]}(i,i) + R_{s[t_2:t_K]}(i,i)\right].$$

Suppose that $s[t_2: t_K]$ is the j-th candidate in the list. The cutoff decision for $s[t_2: t_K]$ is made by:

$$\mathbb{I}^\epsilon_{co}(s[t_2:t_K]) = \begin{cases} \text{Pass,} & \text{if } \tilde{\mathcal{I}}(s[t_2:t_K]) \geq \mathcal{I}_{LBD} - \epsilon \\ \text{fail,} & \text{otherwise,} \end{cases} \tag{12.16}$$

where $\mathcal{I}_{LED}$ is the tightest lower bound on the optimal information reward based on the previous $(j-1)$ solution candidates. ϵ is the relaxation parameter to reduce possibility of rejecting potentially good candidates. Thus, with this heuristic method, the actual mutual information of candidate $\mathbf{s}[t_2:t_K]$ is computed, only if $\mathbb{I}_{co}(\mathbf{s}[t_2:t_K]) = $ pass.

12.4.5 DECOMPOSITION SCHEMES

The proposed cut-off heuristics can reduce the number of computations of matrix determinants. However, the technique still requires repeated calculations of cost-to-go values combinatorially. Thus, a further approximation scheme that breaks down the original problem into a set of small problems is needed to address larger-scale problems.

Receding horizon planning is a common way to decompose a sequential decision making problem along the temporal dimension. Suppose an overall plan has horizon K, that is, it has K steps. The plan can be divided into n_H shorter horizons; a plan for the first horizon consisting of K/n_H waypoints is computed, and the plan for the next horizon is computed conditioned on the measurement selection of the first horizon. This process is repeated through the n_H-th horizon.

In the spatial dimension, the overall UAS fleet (size n_S) is decomposed into n_T teams of size n_S/n_T. Within this team-based decomposition framework, agents are coordinated by a *sequential* decision topology in which each team knows the plans of preceding teams and incorporates the plans into its decision making. Teams make decisions about data for planning by using the conditional covariance matrix conditioned on the preceding teams' plans. For example, a team of sensors $(lm_T + 1)$ to $(l + 1)m_T$ will compute their best control decisions based on the covariance matrices:

$$\text{Cov}(X_{S_Z} | Z_{s_{\{1:lm_T\}}[t_{kmH}+2:t_{(k+1)mH}+1]}, Z_{s[t_2:t_{kmH}+1]}, Z_{s_o}), \quad \text{and}$$

$$\text{Cov}(X_{S_Z} | V, Z_{s_{\{1:lm_T\}}[t_{kmH}+2:t_{(k+1)mH}+1]}, Z_{s[t_2:t_{kmH}+1]}, Z_{s_o})$$

where

$$Z_{s_{\{1:lm_T\}}[t_{kmH}+2:t_{(k+1)mH}+1]} \triangleq \bigcup_{i=1}^{lm_T} Z_{s_i[t_{kmH}+2:t_{(k+1)mH}+1]}.$$

12.5 NUMERICAL TARGETING EXAMPLES

12.5.1 SETUP

A simplified weather forecasting problem is considered for numerical validation of the proposed coordinated targeting algorithm. The nonlinear model utilized for numerical studies is the Lorenz 2003 [29]. It is an idealized chaos model that addresses the multi-scale features of weather dynamics along with the basic aspects of weather systems such as energy dissipation, advection, and external forcing that were captured in the earlier Lorenz 95 model [14].

The Lorenz models have been successfully implemented for the initial verification of numerical weather prediction algorithms [13,14]. In this chapter, the original one-dimensional model [14,29] is extended to two-dimensions to represent the global dynamics of the mid-latitude region (20 to 70 degrees) of the northern hemisphere [26]. The system equations of the Lorenz 2003 model are:

$$\dot{\phi}_{ij} = -\phi_{ij} - \zeta_{i-4,j}\,\zeta_{i-2,j} + \frac{1}{3}\sum_{k\in\{-1,0,1\}}\zeta_{i-2+k,j}\phi_{i+2+k,j}$$

$$-\mu\eta_{i,j-4}\eta_{i,j-2} + \frac{\mu}{3}\sum_{k\in\{-1,0,1\}}\eta_{i,j-2+k}\phi_{i,j+2+k} + \phi_0$$

where $\zeta_{ij} \triangleq (1/3)\Sigma_{k\in\{-1,0,1\}}\phi_{i+k,j}$, $\eta_{ij} \triangleq (1/3)\Sigma_{k\in\{-1,0,1\}}\phi_{i,j+k}$ for $(i,j)\in\{1,2,...,L_i\}\times\{1,2,...,L_j\}$. The state variable ϕ_{ij} denotes a scalar meteorological quantity such as vorticity or temperature [14] at the (i,j)-th grid point. The boundary conditions $\phi_{i+L_i,j} = \phi_{i-L_i,j} = \phi_{i,j}$ and $\phi_{i,0} = \phi_{i,-1} = 3$, $\phi_{i,L_j+1} = 0$ in advection terms are applied to model the mid-latitude area of the northern hemisphere as an annulus. The parameter values are $L_i = 72$, $L_j = 17$, $\mu = 0.66$, and $\phi_0 = 8$. The 1×1 grid size corresponds to 347×347 km in real distance and 1 time unit is equivalent to 5 days in real time.

We tracked the Lorenz 2003 dynamics with an EnSRF incorporating measurements from a fixed observation network of size 186. The distribution of the routine observations network is depicted with asterisks in Figure 12.3. The static network is dense in two portions of the grid space representing land and sparse in the other two portions representing oceans. The routine network takes measurements every 6 hours (0.05 time unit).

The leftmost part consisting of 27 grid points in the right land is selected as the verification region. The verification time was set 3 days after the last targeted observations: $t_V = t_K + 3\text{days}$. The planning window is given by $t_1 = 3\text{hrs}$ and $t_K = 15$ hours

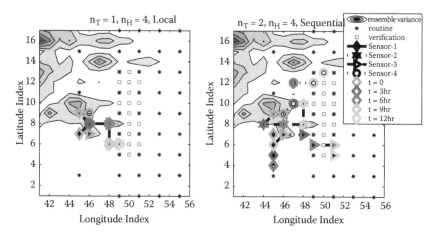

FIGURE 12.3 Targeting solutions for sensors initially close to each other: Different markers represent different sensors. (Source: IEEE© 2011.)

with $K = 5$ (thus $\Delta t = 3$ hours). The analysis ensemble at t_0, $\mathbf{X}^a(t_0)$, with size 1224 is obtained by running an EnSRF incorporating the routine observations for 1000 cycles.

The contour plots in Figure 12.3 show a typical shape of the error variance field with this routine configuration. The measurement noise variance is $R_{routine} = 0.2^2$ for routines and $R_s = 0.02^2$ for additional observations, assuming that high resolution sensors will be equipped on the mobile platforms. Flight speed of the sensor platform is limited by $v_{min} = 1$ grid/timestep (~ 116km/hr) and $v_{max} = 2$ grids/timestep (~ 232km/hr).

Figure 12.3 shows one exemplary targeting solution applying the decomposition scheme with $m_T = 2_s$, $m_H = 4$ and the cutoff parameter $\epsilon = 0.2$. The sensors in the right picture are exploring a relatively large portion of the space. Monte Carlo simulations for similar targeting problems indicate that the average computation time for problems of this size is 1,075 seconds, while the full optimal solution equivalent to $(m_T, m_H) = (4,4)$ cannot be obtained tractably. Regarding the potential suboptimality due to decomposition, the degree of suboptimality has not been quantified (due to the lack of an optimal solution) but compared to random selection, this scheme provides 62% improvement of the resulting mutual information.

12.6 SENSITIVITY ANALYSIS FOR POSSIBLE REAL-SCALE EXTENSION

An ensemble-based estimation framework is a relatively computationally efficient way to address nonlinearity in the dynamics of complex systems. Note that the size of ensemble is still the practical limitation in estimating very large scale dynamic systems such as weather. For example, a typical mesoscale weather dynamic is described by millions of state variables.

High performance computation power is still just enough to create and/or operate hundreds (or even tens of) ensembles for a few-day forecasting of such dynamics. The accuracy of ensemble approximation depends heavily on the size of ensemble. It is crucial to analyze the effect of ensemble size on the solution quality of the targeting approach as the first-step to consider in applying our targeting methodology to real-world-scale weather forecasting problems.

This section provides such an analysis. Note that is makes no strong claims about limitations of the method. Our intention is to present a methodology to quantify and predict impacts.

Before providing detailed analysis, we should first note that the backward ensemble formulation in Reference (11) is not subject to rank deficiency so that it is well defined for nontrivial measurement noise. That might not be the case if the forward approach is taken for an ensemble size smaller than the verification region.

12.6.1 Analysis of Sample Estimation Statistics

Consider an unbiased estimator of log det P for $P \succ 0 \in \mathbb{R}^{n \times n}$:

$$\widehat{\mathcal{H}} = \log \det \left(\frac{m}{2^n} \hat{P} \right) - \prod_{i=0}^{n-1} \psi \left(\frac{m-i}{2} \right) \tag{12.17}$$

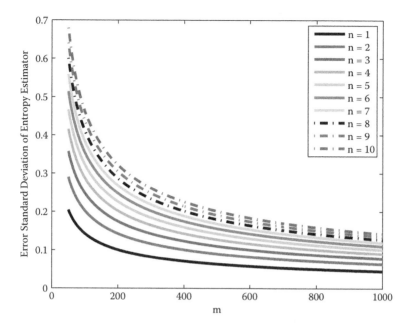

FIGURE 12.4 Standard deviation of $\widehat{\mathcal{H}} - \mathcal{H}$.

where $m = L_E - 1$, which is introduced for notational simplicity in this section. $\hat{P} \equiv \widetilde{\mathbf{X}}\widetilde{\mathbf{X}}^T / m$ is the minimum-variance unbiased estimator of P [30]. The estimation error variance then becomes:

$$\mathbb{E}\left[\left(\widehat{\mathcal{H}} - \mathcal{H}\right)^2\right] = \sum_{i=0}^{n-1} \psi^{(1)}\left(\frac{m-i}{2}\right), \tag{12.18}$$

which depends on the dimension of the random vector n, the sample size L_E, but not the true value of H. Figure 12.4 depicts the error standard deviation for various values of n and m. The plots show that large n and small m lead to large estimation error (on the order of $\sqrt{n/m}$). This dependency of error standard deviation on the order of $\sqrt{n/m}$ will be utilized to determine the impact of limited ensemble size on the performance of ensemble-based targeting in the following section.

12.6.1.1 Range-to-Noise Ratio

The ensemble-based targeting problem must distinguish the best measurement candidate from other suboptimal measurement candidates. An important predictor of the degree of impact that limitation of sample size may exert on solution optimality is the *range-to-noise ratio* of the mutual information values:

$$\text{RNR} = \frac{\sup_s \mathcal{I}_{(s)} - \inf_s \mathcal{I}_{(s)}}{\sup_s \sqrt{\mathbb{E}\left[\left(\widehat{\mathcal{I}}(\mathbf{S}) - \mathcal{I}(\mathbf{S})\right)^2\right]}} \tag{12.19}$$

where $\widehat{\mathcal{I}}(\mathbf{s})$ and $\mathcal{I}(\mathbf{s})$ denote the predicted and the actual mutual information values for Z_s. Utilizing the statistics of the entropy estimation described in the previous section, the *sup* value in the denominator can be obtained without regard to the true values of $\mathcal{I}(\mathbf{s})$. Estimation of mutual information can be treated as estimation and subtraction of the two (prior and posterior) entropies. Since the bias term $\prod_{i=0}^{n-1} \psi\left(\dfrac{m-i}{2}\right)$ is the same for the prior and posterior entropy estimation, the estimation error of the mutual information can be expressed as:

$$
\mathbb{E}\left\{\left[\frac{1}{2}(\widehat{\mathcal{H}}^- - \widehat{\mathcal{H}}^+) - \frac{1}{2}(\mathcal{H}^- - \mathcal{H}^+)\right]^2\right\}
$$

$$
= \frac{1}{4}\mathbb{E}\left[(\widehat{\mathcal{H}}^- - \mathcal{H}^-)^2\right] + \frac{1}{4}\mathbb{E}\left[(\widehat{\mathcal{H}}^+ - \mathcal{H}^+)^2\right] - \frac{1}{2}\mathbb{E}\left[(\widehat{\mathcal{H}}^- - \mathcal{H}^-)(\widehat{\mathcal{H}}^+ - \mathcal{H}^+)\right]
$$

$$
= \frac{1}{2}\sum_{i=0}^{n-1}\psi^{(1)}\left(\frac{m-i}{2}\right) - \frac{1}{2}\mathbb{E}\left[(\widehat{\mathcal{H}}^- - \mathcal{H}^-)(\widehat{\mathcal{H}}^+ - \mathcal{H}^+)\right]
$$

where the minus and plus superscripts denote the prior and posterior, respectively. The cross-correlation term in the final expression is always non-negative; so the estimation error of the mutual information is upper-bounded by:

$$
\mathbb{E}\left[\left(\widehat{\mathcal{I}}(\mathbf{s}) - \mathcal{I}(\mathbf{s})\right)^2\right] \leq \frac{1}{2}\sum_{i=0}^{n-1}\psi^{(1)}\left(\frac{m-i}{2}\right) \triangleq \sigma^2_{m,n}
$$

where equality holds if the prior and posterior entropy estimators are uncorrelated, which corresponds to infinite mutual information. With this upper bound, the RNR can be approximated as:

$$
\mathrm{RNR} \approx \frac{\sup_s \mathcal{I}(\mathbf{s}) - \inf_s \mathcal{I}(\mathbf{s})}{\sigma_{m,n}}. \tag{12.20}
$$

In contrast to the denominator, the numerator of Equation (12.20) is problem-dependent. Moreover, the *sup* and *inf* values cannot be known unless the true covariances (and equivalently true entropies) are known. Regarding the *inf* value, note that $\mathcal{I}(\mathbf{s})$ is lower bounded by zero; therefore, it is reasonable to say that $\inf_s \mathcal{I}(\mathbf{s})$ is a very small positive quantity. This suggests that we can approximate the *inf* value in the numerator of Equation (12.20) as zero. With regard to the *sup* value, since the 95% confident interval estimator of $\mathcal{I}(\mathbf{s})$ is $\widehat{\mathcal{I}}(\mathbf{s}) \pm 2\sigma_{m,n}$, the interval estimate for RNR is:

$$
\widehat{\mathrm{RNR}} = \left[\max\left\{0, \frac{\sup_s \widehat{\mathcal{I}}(\mathbf{s})}{\sigma_{m,n}} - 2\right\}, \frac{\sup_s \widehat{\mathcal{I}}(\mathbf{s})}{\sigma_{m,n}} + 2\right]
$$

with confidence level 95%. The *max* function ensures that RNR is positive. If the objective of computing RNR is to predict whether a *small* RNR would cause significant performance degradation of the targeting, the following one-sided interval estimator can also be used:

$$\overline{RNR} = \left[0, \frac{\sup_s \hat{\mathcal{I}}(s)}{\sigma_{m,n}} + 1.7 \right] \tag{12.21}$$

with 95% confidence level.

12.6.1.2   Probability of Correct Decision

This section considers the probability that ensemble-based targeting provides a true optimal or $(1 - \varepsilon)$-optimal solution, cited as the *probability of correct decision* (PCD) hereafter, for given values of RNR, m, n, and the total number of candidates q. The following are assumed:

1. There are a total of q measurement candidates denoted as $s_1,\ldots,s_q$. Without loss of generality, s_i corresponds to the i-th best targeting solution.
2. The true mutual information values are uniformly distributed over the corresponding range $\mathcal{I}(s_1) - \mathcal{I}(s_q)$. In other words, $\mathcal{I}(s_i) = \mathcal{I}(s_1) - (i-1)\delta$ where $\delta = \dfrac{\mathcal{I}(s_1) - \mathcal{I}(s_q)}{q-1} = RNR \times \dfrac{\sigma_{m,n}}{q-1}$.
3. The estimation error of each mutual information value is distributed with $\mathcal{N}(0, \sigma_{m,n}^2)$.
4. The estimation errors of the mutual information for each measurement candidate are uncorrelated each other. In other words, $\mathbb{E}\left[(\hat{\mathcal{I}}(s_i) - \mathcal{I}(s_i))(\hat{\mathcal{I}}(s_j) - \mathcal{I}(s_j)) \right] = 0, \forall i \neq j$.

Under these assumptions, it can be shown that for $i \leq q - 1$:

$$D_i \triangleq \hat{\mathcal{I}}(s_1) - \hat{\mathcal{I}}(s_i) \sim \mathcal{N}\left((i-1)\delta, 2\sigma_{m,n}^2 \right) \tag{12.22}$$

and

$$\mathrm{Cov}\,(D_i, D_j) = \mathbb{E}[(D_i - (i-1)\delta)(D_j - (j-1)\delta)] = \sigma_{m,n}^2, \forall i \neq j. \tag{12.23}$$

Since PCD can be interpreted as the probability that ensemble-based targeting will declare s_1 to be the best candidate, PCD can be written in terms of D_is as PCD = $\mathrm{Prob}[D_i > 0, \forall i]$. Using Equations (12.22) and (12.23), the PCD can be computed:

$$PCD = \int_{-\infty}^{RNR} \cdots \int_{-\infty}^{\frac{i}{q-1}RNR} \cdots \int_{-\infty}^{\frac{RNR}{q-1}} f_{\mathcal{N}(0,\Sigma)}(x_1,\ldots,x_{q-1}) dx_1 \ldots dx_{q-1} \tag{12.24}$$

where $f_{\mathcal{N}(0,\Sigma)}$ is the pdf of the zero-mean multivariate Gaussian distribution with the covariance matrix of $\Sigma = I_{q-1} + \mathbf{1}_{q-1} \otimes \mathbf{1}_{q-1}^T$ where I_{q-1} denotes the $(q-1) \times (q-1)$ identity matrix, $\mathbf{1}_{q-1}$ is the $(q-1)$-dimension column vector with every element being unity, and $\otimes$ denotes the Kronecker product. That is, all the diagonal elements of Σ_q are 2s, while all the off-diagonal elements are 1s.

Note that the PCD is expressed as a cdf of a $(q-1)$-dimensional normal distribution. In the special case of $q = 2$,

$$PCD_{q=2} = \Phi\left(\frac{RNR}{\sqrt{2}}\right)$$

where $\Phi(\cdot)$ is the cdf of the standard normal distribution. Figure 12.5 shows how PCD changes with q and RNR. The plot shows that PCD increases monotonically with respect to RNR, while it decreases monotonically with respect to q. The dependency of PCD on q is crucial because q is a very large number in practice; recall that $q \sim \mathcal{O}\left(\begin{pmatrix} N \\ n \end{pmatrix}\right)$. Thus, PCD can be meaninglessly small for a large-scale selection problem. In addition, calculating PCD for such large q is computationally very expensive because it requires a cdf evaluation of a large-dimensional normal distribution. For this reason, for a large q case, this work suggests use of the probability of ϵ-correct decision (ϵ-PCD) defined as:

$$\epsilon\text{-PCD} = \text{Prob}\left[\bigcup_{i=1}^{\lfloor \epsilon q \rfloor} \left(\hat{\mathcal{I}}(\mathbf{s}_i)\right) > \hat{\mathcal{I}}(\mathbf{s}_j), \forall j \neq i\right], \tag{12.25}$$

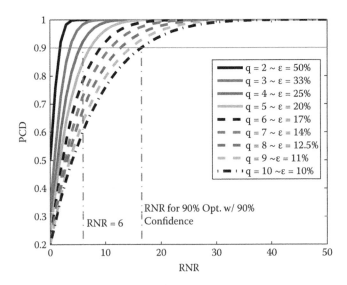

FIGURE 12.5 Probability of correct decision for targeting with q candidates.

since it can still be used as an indicator of the impact of limited sample size on the degree of optimality and can be computed tractably. By the symmetry of the distribution of the true mutual information values, the lower bound of this ϵ-PCD can be computed by:

$$\text{PCD}_{\lfloor 1/\epsilon \rfloor} \leq \epsilon\text{-PCD}, \tag{12.26}$$

where equality holds if $\lfloor \epsilon q \rfloor$ and $\lfloor 1/\epsilon \rfloor$ are integers. In other words, if dividing q candidates into $1/\epsilon$ groups such that the i-th group consists of $s_{(i-1)\epsilon q+1}$ through $s_{i\epsilon q}$, the decision of declaring one of the candidates in the first group to be the optimal solution is equivalent to distinguishing the best among $1/\epsilon$ candidates.

Figure 12.5 can be used to interpret the relation between RNR and $\lceil q \rceil$-PCD. For instance, the graph of PCD for $q = 10$ represents the relation between RNR and 10% PCD for any size targeting problem. The dotted line indicates the RNR value above which 10% PCD is greater than 90%, which is 16.5. *in order to have a 90% optimal targeting solution with a confidence level of 90%, RNR should be greater than 16.5.* The one-sided interval estimator of RNR in Equation (2.21) implies that:

$$\sup_{\mathbf{s}} \widehat{\mathcal{I}}(\mathbf{s}) > 18.2\sigma_{m,n} \approx 18.2\sqrt{n/m} \tag{12.27}$$

for the same qualification with 95% confidence level. The last approximate expression comes from $\sigma_{m,n} \approx \sqrt{n/m}$ for a small n. In addition, the relation in Figure 12.5 can be utilized to determine the level of performance for a given RNR. For instance, as the curve for $q = 4$ crosses the horizontal line of PCD $= 0.9$ when the corresponding RNR is smaller than 6, it can be inferred that RNR $= 6$ guarantees 75% optimality with 90% confidence.

Figure 12.6 depicts the RNR values for $m \in [100,1000]$ and for $n \in [1,5]$. Note that for a given n, RNR decreases as m increases and for a given m, it increases as n increases. For $n = 1$, the requirement of RNR > 16.5 that achieves 90% optimality with 90% confidence is not satisfied even with $m = 1000$, while $m = 400$ meets the same requirement for $n = 5$. Dependency of RNR on m simply indicates that $\sigma_{m,n}$ is an increasing function of m for fixed n. The increasing tendency of RNR with respect to n occurs because the optimal mutual information value grows faster than $\mathcal{O}(\sqrt{n})$.

Since RNR $\approx \dfrac{\sup_{\mathbf{s}} \mathcal{I}(\mathbf{s})}{\sqrt{n/m}}$ for small n, RNR becomes an increasing function of n if the *sup* value in the numerator grows faster than $\mathcal{O}(\sqrt{n})$, which is the case for the Lorenz 95 example cited in this work. Also, seeing that the marginal increment of RNR diminishes as n increases, it is conceivable that there exists a threshold $\bar{n}$ over which increasing n no longer improves RNR.

For the numerical targeting examples in Section 125, the optimality of each decomposed problem was not harmed by the considered ensemble size $L_E = 1024$. However, for real weather forecasting for which fewer than 100 ensembles are available even with high-performance computing capability, the analysis in this section can used to assess and predict the confidence levels of targeting decisions although it cannot eliminate the effects.

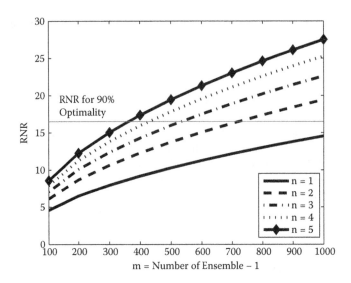

FIGURE 12.6 Range-to-noise ratios with differing *m* and *n*.

12.7 CONCLUSIONS

This chapter presented targeting methodologies for coordinating multiple sensor platforms with constrained mobility. Techniques to efficiently incorporate constraints and further improve computational efficiency were proposed by exploiting the backward formulation for quantifying mutual information. Numerical studies verified the performance and computational tractability of the proposed methods.

This chapter also quantified the impact of limited ensemble size on the performance variations of the targeted solutions in an analytical way. The concepts of range-to-noise ratio and probability of correct decision were proposed to derive mathematical relations between ensemble size and solution sensitivity.

ACKNOWLEDGMENTS

This work was funded in part by CNS-0540331 as part of the National Science Foundation DDDAS program with Dr. Frederica Darema as the program manager and in part by the Basic Science Research Program through the National Research Foundation of Korea funded by the Ministry of Education, Science and Technology (2010-0025484).

REFERENCES

1. A.E. MacDonald. 2005. A global profiling system for improved weather and climate prediction. *Bulletin of American Meteorological Society*, 86, 1747–1764.
2. M. Aslaksen, T.A. Jacobs, J.D. Sellars et al. 2009. Altair unmanned aircraft system achieves demonstration goals. *EOS Transactions on AGU*, 87, 197–201.
3. J.C. Rubio. 2004. Long range evolution-based path planning for UAVs through realistic weather environments. Master's thesis, University of Washington.

4. J. C. Rubio, J. Vagners, and R. Rysdyk. 2004. Adaptive path planning for autonomous UAV oceanic search missions. *AIAA Intelligent Systems Technical Conference.*
5. J. Elston and E.W. Frew. 2010. Unmanned aircraft guidance for penetration of pre-tornadic storms. *AIAA Journal of Guidance, Control, and Dynamics*, 33, 99–107.
6. J. Elston, M. Stachura, E.W. Frew et al. 2009. Toward model free atmospheric sensing by aerial robot networks in strong wind fields. In *Proceedings of IEEE International Conference on Robotics and Automation.*
7. J.S. Elston, J. Roadman, M. Stachura et al. 2011. The Tempest unmanned aircraft system for in situ observations of tornadic supercells. *Journal of Field Robotics*, 28, 461–483.
8. D.N. Daescu and I.M. Navon. 2004. Adaptive observations in the context of 4D-var data assimilation. *Meteorology and Atmospheric Physics*, 85, 205–226.
9. G. Evensen. 2004. Sampling strategies and square root analysis schemes for EnKF. *Ocean Dynamics*, 54, 539–560.
10. T.M. Hamill and C. Snyder. 2002. Using improved background error covariances from an ensemble Kalman filter for adaptive observations. *Monthly Weather Review*, 130, 1552–1572.
11. J.A. Hansen and L.A. Smith. 2000. The role of operational constraints in selecting supplementary observation. *Journal of Atmospheric Sciences*, 57, 2859–2871.
12. C. Kohl and D. Stammer. 2004. Optimal observations for variational data assimilation. *Journal of Physical Oceanography*, 34, 529–542.
13. M. Leutbecher. 2003. A reduced rank estimate of forecast error variance changes due to intermittent modifications of the observing network. *Journal of Atmospheric Sciences*, 60, 729–742.
14. E. Lorenz and K. Emanuel. 1998. Optimal sites for supplementary weather observations: simulation with a small model. *Journal of Atmospheric Sciences*, 55, 399–414, 1998.
15. S. Majumdar, C. Bishop, B. Etherton et al. 2002. Adaptive sampling with the ensemble transform Kalman filter II. Field programming implementation. *Monthly Weather Review*, 130, 1356–1369.
16. T. Palmer, R. Gelaro, J. Barkmeijer et al. 1998. Singular vectors, metrics, and adaptive observations. *Journal of Atmospheric Sciences*, 55, 633–653.
17. H.L. Choi and J.P. How. 2010. Continuous trajectory planning of mobile sensors for informative forecasting. *Automatica*, 46, 1266–1275.
18. S. Park, H.L. Choi, N. Roy et al. 2010. Learning the covariance dynamics of a large-scale environment for informative path planning of unmanned aerial vehicle sensors. *International Journal of Aeronautical and Space Sciences*, 11, 326–337.
19. H.L. Choi and J.P. How. 2011. Coordinated targeting of mobile sensor networks for ensemble forecast improvement. *IEEE Sensors Journal*, 11, 621–633.
20. H.L. Choi and J.P. How. 2011. Efficient targeting of sensor networks for large-scale systems. *IEEE Transactions on Control Systems Technology*, 19, 1569–1577.
21. H.L. Choi, J.P. How, and J.A. Hansen. 2008. Algorithm and sensitivity analysis of information-theoretic ensemble-based observation targeting. *American Meteorological Society 19th Annual Conference on Probability and Statistics.*
22. J. Whitaker and H. Hamill. 2002. Ensemble data assimilation without perturbed observations. *Monthly Weather Review*, 130, 1913–1924.
23. G. Evensen. 2003. The ensemble Kalman filter: theoretical formulation and practical implementation. *Ocean Dynamics*, 53, 343–367.
24. T.M. Hamill, J.S. Whitaker, and C. Snyder. 2001. Distance-dependent filtering of background error covariance estimates in an ensemble Kalman filter. *Monthly Weather Review*, 129, 2776–2790.
25. R. Buizza and T.N. Palmer. 1998. Impact of ensemble size on ensemble prediction. *Monthly Weather Review*, 126, 2503–2518.

26. J.A. Hansen. n.d. Personal communication from Naval Research Laboratory, Monterey, CA.

27. T. Cover and J. Thomas. 1991. *Elements of Information Theory.* New York: John Wiley & Sons.

28. B. Andersen, J. Gunnels, F. Gustavson et al. 2002. A recursive formulation of the inversion of symmetric positive definite matrices in packed storage data format. *Lecture Notes in Computer Science*, 2367, 287–296.

29. E.N. Lorenz. 2005. Designing chaotic models. *Journal of Atmospheric Sciences*, 62, 1574–1587.

30. N. Misra, H. Singh, and E. Demchuck. 2005. Estimation of the entropy of a multivariate normal distribution. *Journal of Multivariate Analysis*, 92, 324–342.

Index

Printed and bound by CPI Group (UK) Ltd, Croydon, CR0 4YY

18/10/2024

01776266-0001